Lecture Notes in Earth Sciences

Vol. 1: Sedimentary and Evolutionary Cycles. Edited by U. Bayer and A. Seilacher. VI, 465 pages. 1985.

Vol. 2: U. Bayer, Pattern Recognition Problems in Geology and Paleontology. VII, 229 pages. 1985.

Vol. 3: Th. Aigner, Storm Depositional Systems. VIII, 174 pages. 1985.

Vol. 4: Aspects of Fluvial Sedimentation in the Lower Triassic Buntsandstein of Europe. Edited by D. Mader. VIII, 626 pages. 1985.

Vol. 5: Paleogeothermics. Edited by G. Buntebarth and L. Stegena. II, 234 pages. 1986.

Vol. 6: W. Ricken, Diagenetic Bedding. X, 210 pages. 1986.

Vol. 7: Mathematical and Numerical Techniques in Physical Geodesy, Edited by H. Sünkel. IX, 548 pages. 1986.

Vol. 8: Global Bio-Events. Edited by O.H. Walliser. IX, 442 pages. 1986.

Vol. 9: G. Gerdes, W.E. Krumbein, Biolaminated Deposits. IX, 183 pages. 1987.

Vol. 10: T.M. Peryt (Ed.), The Zechstein Facies in Europe. V, 272 pages. 1987.

Vol. 11: L. Landner (Ed.), Contamination of the Environment. Proceedings, 1986. VII, 190 pages. 1987.

Vol. 12: S. Turner (Ed.), Applied Geodesy. VIII, 393 pages. 1987.

Vol. 13: T.M. Peryt (Ed.), Evaporite Basins. V, 188 pages. 1987.

Vol. 14: N. Cristescu, H.I. Ene (Eds.), Rock and Soil Rheology. VIII, 289 pages. 1988.

Vol. 15: V.H. Jacobshagen (Ed.), The Atlas System of Morocco. VI, 499 pages. 1988.

Vol. 16: H. Wanner, U. Siegenthaler (Eds.), Long and Short Term Variability of Climate. VII, 175 pages. 1988.

Vol. 17: H. Bahlburg, Ch. Breitkreuz, P. Giese (Eds.), The Southern Central Andes. VIII, 261 pages. 1988.

Vol. 18: N.M.S. Rock, Numerical Geology. XI, 427 pages. 1988.

Vol. 19: E. Groten, R. Strauß (Eds.), GPS-Techniques Applied to Geodesy and Surveying. XVII, 532 pages. 1988.

Vol. 20: P. Baccini (Ed.), The Landfill. IX, 439 pages. 1989.

Vol. 21: U. Förstner, Contaminated Sediments. V, 157 pages. 1989.

Vol. 22: I.I. Mueller, S. Zerbini (Eds.), The Interdisciplinary Role of Space Geodesy. XV, 300 pages. 1989.

Vol. 23: K.B. Föllmi, Evolution of the Mid-Cretaceous Triad. VII, 153 pages. 1989.

Vol. 24: B. Knipping, Basalt Intrusions in Evaporites. VI, 132 pages. 1989.

Vol. 25: F. Sansò, R. Rummel (Eds.), Theory of Satellite Geodesy and Gravity Field Theory. XII, 491 pages. 1989.

Vol. 26: R.D. Stoll, Sediment Acoustics. V, 155 pages. 1989.

Vol. 27: G.-P. Merkler, H. Militzer, H. Hötzl, H. Armbruster, J. Brauns (Eds.), Detection of Subsurface Flow Phenomena. IX, 514 pages. 1989.

Vol. 28: V. Mosbrugger, The Tree Habit in Land Plants. V, 161 pages. 1990.

Vol. 29: F.K. Brunner, C. Rizos (Eds.), Developments in Four-Dimensional Geodesy. X, 264 pages. 1990.

ctd. on inside back cover

Lecture Notes in Earth Sciences

Edited by Somdev Bhattacharji, Gerald M. Friedman, Horst J. Neugebauer and Adolf Seilacher

31

Karl-Rudolf Koch

Bayesian Inference with Geodetic Applications

Springer-Verlag
Berlin Heidelberg GmbH

Author
Professor Karl-Rudolf Koch
Institute of Theoretical Geodesy, University of Bonn
Nussallee 17, D-5300 Bonn, FRG

ISBN 978-3-540-53080-0 ISBN 978-3-540-46601-7 (eBook)
DOI 10.1007/978-3-540-46601-7

Originally published by Springer-Verlag Berlin Heidelberg New York in 1990
2132/3140-543210 – Printed on acid-free paper

Preface

There are problems, when applying statistical inference to the analysis of data, which are not readily solved by the inferential methods of the standard statistical techniques. One example is the computation of confidence intervals for variance components or for functions of variance components. Another example is the statistical inference on the random parameters of the mixed model of the standard statistical techniques or the inference on parameters of nonlinear models. Bayesian analysis gives answers to these problems.

The advantage of the Bayesian approach is its conceptual simplicity. It is based on Bayes' theorem only. In general, the posterior distribution for the unknown parameters following from Bayes' theorem can be readily written down. The statistical inference is then solved by this distribution. Often the posterior distribution cannot be integrated analytically. However, this is not a serious drawback, since efficient methods exist for the numerical integration.

The results of the standard statistical techniques concerning the linear models can also be derived by the Bayesian inference. These techniques may therefore be considered as special cases of the Bayesian analysis. Thus, the Bayesian inference is more general.

Linear models and models closely related to linear models will be assumed for the analysis of the observations which contain the information on the unknown parameters of the models. The models, which are presented, are well suited for a variety of tasks connected with the evaluation of data. When applications are considered, data will be analyzed which have been taken to solve problems of surveying engineering. This does not mean, of course, that the applications are restricted to geodesy. Bayesian statistics may be applied wherever data need to be evaluated, for instance in geophysics.

After an introduction the basic concepts of Bayesian inference are presented in Chapter 2. Bayes' theorem is derived and the introduction of prior information for the unknown parameters is discussed. Estimates of the unknown parameters, of confidence regions and the testing of hypotheses are derived and the predictive analysis is treated. Finally techniques for the numerical integration of the integrals are presented which have to be solved for the statistical inference.

Chapter 3 introduces models to analyze data for the statistical inference on the unknown parameters and deals with special applications. First the linear model is presented with noninformative and informative priors for the unknown parameters. The agreement with the results of the standard statistical techniques is pointed out. Furthermore, the prediction of data and the linear model not of full rank are discussed. A method for identifying a model is presented and a less sensitive hypothesis test for the standard statistical techniques is derived. The Kalman-Bucy filter for estimating unknown parameters of linear dynamic systems is also given.

Nonlinear models are introduced and as an example the fit of a straight line is treated. The resulting posterior distribution for the unknown parameters is analytically not tractable, so that numerical methods have to be applied for the statistical inference.

In contrast to the standard statistical techniques, the Bayesian analysis for mixed models does not discriminate between fixed and random parameters, it distinguishes the parameters according to their prior information. The Bayesian inference on the parameters, which correspond to the random parameters of the mixed model of the standard statistical techniques, is therefore readily accomplished.

Noninformative priors of the variance and covariance components are derived for the linear model with unknown variance and covariance components. In addition, informative priors are given. Again, the resulting posterior distributions are analytically not tractable, so that numerical methods have to be applied for the Bayesian inference.

The problem of classification is solved by applying the Bayes rule, i.e. the posterior expected loss computed by the predictive density function of the observations is minimized.

Robust estimates of the standard statistical techniques, which are maximum likelihood type estimates, the so-called M-estimates, may also be derived by Bayesian inference. But this approach not only leads to the M-estimates, but also any inferential problem for the parameters may be solved.

Finally, the reconstruction of digital images is discussed. Numerous methods exist for the analysis of digital images. The Bayesian approach unites some of them and gives them a common theoretical foundation. This is due to the flexibility by which prior information for the unknown parameters can be introduced.

It is assumed that the reader has a basic knowledge of the standard statistical techniques. Whenever these results are needed, for easy reference the appropriate page of the book "Parameter Estimation and Hypothesis Testing in Linear Models" by the author (Koch 1988a) is cited. Of course, any other textbook on statistical techniques can serve this purpose.

To easily recognize the end of an example or a proof, it is marked by a Δ or a $\square$, respectively.

I want to thank all colleagues and students who contributed to this book. In particular, I thank Mr. Andreas Busch, Dipl.-Ing., for his suggestions. I also convey my thanks to Mrs. Karin Bauer, who prepared the copy of the book. The assistance of the Springer-Verlag in checking the English text is gratefully acknowledged. The responsibility of errors, of course, remains with the author.

Bonn, June 1990 Karl-Rudolf Koch

Contents

1 Introduction

Bayesian inference, despite its conceptual simplicity, has its opponents. First, it is considered subjective, because it requires the introduction of prior information. However, the use of noninformative priors, also called vague priors, leads to results which are equivalent to the results of the standard statistical techniques. In addition, informative priors can be derived using the concept of maximum entropy, thus making them, except for the information which is conveyed, as noninformative as possible.

Another point of debate is the notion of the unknown parameter in Bayesian analysis. It is defined as a random variable, since a prior and a posterior distribution are associated with it. This does not mean, however, that the unknown parameter cannot represent a constant, like the velocity of light. The prior and posterior density function of the unknown parameter describe the state of knowledge on the parameter. The probability derived from these distributions need not be interpreted as a frequency, that means a measurable quantity. The probability merely represents the state of knowledge and may be considered a subjective quantity. Thus, the unknown parameter can very well mean a constant. For further discussions on the differences between the Bayesian approach and the standard statistical techniques, see Jaynes (1986).

As mentioned in the Preface, geodetic applications, i.e. data analysis for surveying engineering, will be mostly considered. Pope in Dillinger et al. (1971) was the first geodesist to discuss Bayesian inference in order to interpret confidence regions computed for the determination of focal mechanisms. It was Bossler (1972), who introduced Bayesian methods into geodesy by investigating the Bayesian estimates of unknown parameters in linear models with respect to geodetic applications. In particular he dealt with the estimate of the variance of unit weight in the case of prior information. He drew attention to Theil's estimator (Theil 1963), which was later applied for an adjustment of a geodetic network (Bossler and Hanson 1980) and modified for a simpler computation (Harvey 1987). An approximation to the Bayesian estimate of the square root of the variance of unit weight in the linear model was derived by Schaffrin (1987).

Riesmeier (1984) dealt with the test of hypotheses formulated as inequality constraints and developed less sensitive hypothesis tests by modifying the standard statistical techniques (Koch 1988a, p. 310, 319). This method was extended from the univariate model for estimating parameters to the multivariate model by Koch and Riesmeier (1985) and applied to the deformation analysis by Koch (1984, 1985).

Bayesian inference for variance components with the aim of deriving confidence intervals was presented based on noninformative priors by Koch (1987) and informative priors by Koch (1988b). The resulting posterior density functions for the variance components are analytically not tractable, so that the density has to be integrated numerically (Koch 1989). Bayes estimates of variance components for a special model are presented in Ziqiang (1990).

2 Basic Concepts

The starting point of Bayesian inference is Bayes' theorem, which will be derived first. When working with Bayes' theorem the question arises, which prior density function should be associated with the unknown parameters. This, of course, depends on the kind of prior information, and different alternatives for incorporating this information are discussed. Based on the posterior density function derived for the unknown parameters from Bayes' theorem, the estimates of the parameters and their confidence regions are then obtained and methods for testing hypotheses are discussed. The distributions of predicted observations are derived and methods of numerical integration are presented, since integrals have to be solved which are analytically not tractable.

21 Bayes' Theorem

211 Derivation

Let $\mathbf{y}$ be a random vector defined by digital registrations of the outcomes of a random experiment. The values which this random vector takes on or its realizations are called *observations*, *measurements*, or *data*. They are collected in a vector, which is also denoted by $\mathbf{y}$, to simplify the notation. Let the probability density function of the random vector $\mathbf{y}$ be dependent on unknown *parameters*. They are defined as random variables and collected in the random vector $\boldsymbol{\theta}$. The random parameter vector $\boldsymbol{\theta}$ has the probability density function $p(\boldsymbol{\theta})$, where again for the sake of simple notation $\boldsymbol{\theta}$ denotes the vector of realizations. We assume $\boldsymbol{\theta} \subset \Theta$, with Θ being the set of admissible parameter vectors or the *parameter space*. The density function of the random vector $\mathbf{y}$ is introduced as the conditional density function $p(\mathbf{y}|\boldsymbol{\theta})$ given that the random parameter vector $\boldsymbol{\theta}$ is equal to its realization.

Bayes' Theorem: The probability density function $p(\boldsymbol{\theta}|\mathbf{y})$ of the parameters $\boldsymbol{\theta}$ given the observations $\mathbf{y}$ is obtained from the density $p(\boldsymbol{\theta})$ of the parameters $\boldsymbol{\theta}$ and the density $p(\mathbf{y}|\boldsymbol{\theta})$ of the observations $\mathbf{y}$ given the parameters $\boldsymbol{\theta}$ by

$$p(\boldsymbol{\theta}|\mathbf{y}) \propto p(\boldsymbol{\theta})p(\mathbf{y}|\boldsymbol{\theta}),$$

where $\propto$ denotes proportionality. (211.1)

Proof: The conditional density function $p(\mathbf{y}|\boldsymbol{\theta})$ is defined by (Koch 1988a, p.107)

$$p(\mathbf{y}|\boldsymbol{\theta}) = \frac{p(\mathbf{y},\boldsymbol{\theta})}{p(\boldsymbol{\theta})}, \tag{211.2}$$

where $p(\mathbf{y},\boldsymbol{\theta})$ denotes the joint density function of $\mathbf{y}$ and $\boldsymbol{\theta}$. Solving for it and applying (211.2) once more leads to

$$\begin{aligned} p(\mathbf{y},\boldsymbol{\theta}) &= p(\boldsymbol{\theta})p(\mathbf{y}|\boldsymbol{\theta}) \\ &= p(\mathbf{y})p(\boldsymbol{\theta}|\mathbf{y}) \end{aligned}$$

or by equating the right-hand sides to

$$p(\boldsymbol{\theta}|\mathbf{y}) = \frac{p(\boldsymbol{\theta})p(\mathbf{y}|\boldsymbol{\theta})}{p(\mathbf{y})} \quad \text{for} \quad p(\mathbf{y}) > 0. \tag{211.3}$$

This proves the theorem, since the observations $\mathbf{y}$ are given, so that $p(\mathbf{y})$ is a constant. □

The density $p(\mathbf{y})$ in (211.3) may be obtained from the joint density $p(\mathbf{y},\boldsymbol{\theta})$ as a marginal density by (Koch 1988a, p.105)

$$p(\mathbf{y}) = \int_{\Theta} p(\mathbf{y},\boldsymbol{\theta})d\boldsymbol{\theta}$$
$$= \int_{\Theta} p(\boldsymbol{\theta})p(\mathbf{y}|\boldsymbol{\theta})d\boldsymbol{\theta},$$

where the domain of the integration is the parameter space Θ. We therefore obtain the alternative formulation of Bayes' theorem

$$p(\boldsymbol{\theta}|\mathbf{y}) = cp(\boldsymbol{\theta})p(\mathbf{y}|\boldsymbol{\theta}) \tag{211.4}$$

with

$$c = 1/(\int_{\Theta} p(\boldsymbol{\theta})p(\mathbf{y}|\boldsymbol{\theta})d\boldsymbol{\theta}), \tag{211.5}$$

where c is a normalization constant, which ensures that the density function $p(\boldsymbol{\theta}|\mathbf{y})$ fulfills the conditions (Koch 1988a, p.104)

$$p(\boldsymbol{\theta}|\mathbf{y}) \geq 0 \quad \text{and} \quad \int_{\Theta} p(\boldsymbol{\theta}|\mathbf{y})d\boldsymbol{\theta} = 1. \tag{211.6}$$

The density function $p(\boldsymbol{\theta})$ of the parameters $\boldsymbol{\theta}$ in (211.1) summarizes what is known about the parameters $\boldsymbol{\theta}$, before the data $\mathbf{y}$ have been taken. It is therefore called the *prior density* or the *prior distribution* of the parameters $\boldsymbol{\theta}$. With the introduction of the observations $\mathbf{y}$ the density $p(\boldsymbol{\theta}|\mathbf{y})$ for $\boldsymbol{\theta}$ is obtained. It is called the *posterior density* or the *posterior distribution* of the parameters $\boldsymbol{\theta}$. The information on the parameters $\boldsymbol{\theta}$ coming from the data $\mathbf{y}$ enters through the density $p(\mathbf{y}|\boldsymbol{\theta})$. Since $\mathbf{y}$ is given, this density may be viewed not as a function of $\mathbf{y}$ but as a function of the parameters $\boldsymbol{\theta}$. The density $p(\mathbf{y}|\boldsymbol{\theta})$ is therefore called the *likelihood function*. It is often denoted by $l(\boldsymbol{\theta}|\mathbf{y})$ in order to indicate that it is a function of $\boldsymbol{\theta}$. We may therefore express Bayes' theorem by

posterior density function $\propto$ prior density function $\times$ likelihood function.

The observations modify through the likelihood function the prior knowledge of the parameters, thus leading to the posterior density function of the parameters.

The posterior distribution solves the problem of the statistical inference. By means of this distribution the estimates of the unknown parameters and their confidence regions are computed and the probabilities are determined to decide whether to accept or reject hypotheses for the unknown parameters.

Example 1: Suppose we have n independent observations $\mathbf{y}=[y_1,\ldots,y_n]'$, for instance of the length of a straight line or of an angle. Let each observation be normally distributed with the unknown expected value μ und the known variance σ^2. We want to determine the posterior density function $p(\mu|\mathbf{y})$ for the unknown paramter μ with the prior density function $p(\mu)$ of μ resulting from a normal distribution. With (211.1) we obtain

$$p(\mu|\mathbf{y}) \propto p(\mu)p(\mathbf{y}|\mu). \tag{211.7}$$

As required, $p(\mu)$ is represented by the normal distribution (A11.1)

$$p(\mu) = \frac{1}{\sqrt{2\pi}\,\sigma_\mu} \exp[-\frac{1}{2\sigma_\mu^2}(\mu-\mu_p)^2]. \tag{211.8}$$

The prior information on the expected value and the variance of the parameter μ is therefore given by μ_p and σ_μ^2.

Each observation y_i is normally distributed, hence

$$p(y_i|\mu) = \frac{1}{\sqrt{2\pi}\,\sigma} \exp[-\frac{1}{2\sigma^2}(y_i-\mu)^2]. \tag{211.9}$$

Since the observations y_i are independent, their joint distribution follows from the product of the individual distributions (Koch 1988a, p.107), so that the likelihood function is given by

$$p(\mathbf{y}|\mu) = (2\pi\sigma^2)^{-n/2} \exp[-\frac{1}{2\sigma^2}\sum_{i=1}^{n}(y_i-\mu)^2]. \tag{211.10}$$

With the estimate $\hat{\mu}$ of μ (Koch 1988a, p.193)

$$\hat{\mu} = \frac{1}{n}\sum_{i=1}^{n} y_i \tag{211.11}$$

and with the estimate $\hat{\sigma}^2$ of σ^2

$$\hat{\sigma}^2 = \frac{1}{n-1}\sum_{i=1}^{n}(y_i-\hat{\mu})^2 \tag{211.12}$$

we obtain for the exponent

$$\sum_{i=1}^{n}(y_i-\mu)^2 = \sum_{i=1}^{n}((y_i-\hat{\mu})-(\mu-\hat{\mu}))^2$$
$$= \sum_{i=1}^{n}(y_i-\hat{\mu})^2 + n(\mu-\hat{\mu})^2, \tag{211.13}$$

since $\sum_{i=1}^{n}(y_i-\hat{\mu})(\mu-\hat{\mu})=0$ because of $\sum_{i=1}^{n}(y_i-\hat{\mu})=0$. Thus, we find instead of (211.10)

$$p(\mathbf{y}|\mu) = (2\pi\sigma^2)^{-n/2} \exp[-\frac{1}{2\sigma^2}((n-1)\hat{\sigma}^2+n(\mu-\hat{\mu})^2)].$$

The likelihood function can be factored into a term only depending on $\mathbf{y}$ through $\hat{\sigma}^2$ and into a term depending on μ and $\hat{\mu}$

$$p(\mathbf{y}|\mu) = (2\pi\sigma^2)^{-n/2} \exp[-\frac{1}{2\sigma^2}(n-1)\hat{\sigma}^2]\exp[-\frac{n}{2\sigma^2}(\mu-\hat{\mu})^2]. \qquad (211.14)$$

When substituting $p(\mathbf{y}|\mu)$ in (211.7) to derive the posterior density $p(\mu|\mathbf{y})$ of μ given $\mathbf{y}$, the first term is constant. Hence, by neglecting the constants also in (211.8) we find

$$p(\mu|\mathbf{y}) \propto \exp[-\frac{1}{2}(\frac{(\mu-\mu_p)^2}{\sigma_\mu^2} + \frac{n}{\sigma^2}(\mu-\hat{\mu})^2)]. \qquad (211.15)$$

On completing the squares of the exponent we find

$$\frac{(\mu-\mu_p)^2}{\sigma_\mu^2} + \frac{n}{\sigma^2}(\mu-\hat{\mu})^2 = \frac{1}{\sigma_\mu^2\sigma^2}(\mu^2(n\sigma_\mu^2+\sigma^2) - 2\mu(n\hat{\mu}\sigma_\mu^2+\mu_p\sigma^2) + n\hat{\mu}^2\sigma_\mu^2 + \mu_p^2\sigma^2)$$

$$= \frac{\sigma_\mu^2+\sigma^2/n}{\sigma_\mu^2\sigma^2/n}(\mu^2 - 2\mu\frac{\hat{\mu}\sigma_\mu^2+\mu_p\sigma^2/n}{\sigma_\mu^2+\sigma^2/n} + (\frac{\hat{\mu}\sigma_\mu^2+\mu_p\sigma^2/n}{\sigma_\mu^2+\sigma^2/n})^2$$

$$-(\frac{\hat{\mu}\sigma_\mu^2+\mu_p\sigma^2/n}{\sigma_\mu^2+\sigma^2/n})^2 + \frac{\hat{\mu}^2\sigma_\mu^2+\mu_p^2\sigma^2/n}{\sigma_\mu^2+\sigma^2/n}). \qquad (211.16)$$

The last two terms on the right-hand side are constant and may be neglected. We therefore obtain the posterior density function $p(\mu|\mathbf{y})$ of μ

$$p(\mu|\mathbf{y}) \propto \exp[-\frac{1}{2}\frac{\sigma_\mu^2+\sigma^2/n}{\sigma_\mu^2\sigma^2/n}(\mu - \frac{\hat{\mu}\sigma_\mu^2+\mu_p\sigma^2/n}{\sigma_\mu^2+\sigma^2/n})^2]. \qquad (211.17)$$

Now we recognize with (A11.1) and (A11.3) that $p(\mu|\mathbf{y})$ is given by the normal distribution

$$\mu|\mathbf{y} \sim N(E(\mu),\ V(\mu))$$

with the expected value

$$E(\mu) = \frac{\hat{\mu}\sigma_\mu^2+\mu_p\sigma^2/n}{\sigma_\mu^2+\sigma^2/n} = \frac{\hat{\mu}(\sigma^2/n)^{-1}+\mu_p(\sigma_\mu^2)^{-1}}{(\sigma^2/n)^{-1}+(\sigma_\mu^2)^{-1}} \qquad (211.18)$$

and the variance

$$V(\mu) = \frac{\sigma_\mu^2\sigma^2/n}{\sigma_\mu^2+\sigma^2/n} = \frac{1}{(\sigma^2/n)^{-1}+(\sigma_\mu^2)^{-1}}. \qquad (211.19)$$

The variance of the mean $\hat{\mu}$ is σ^2/n (Koch 1988a, p.193) and the variance of the prior information for μ is σ_μ^2. The reciprocals of the variances $(\sigma^2/n)^{-1}$ and $(\sigma_\mu^2)^{-1}$ are defined as weights (Koch 1988a, p.120). They are introduced into (211.18) and (211.19) by multiplying numerator and denominator by $1/(\sigma^2_\mu \sigma^2/n)$. The expected value $E(\mu)$ of the unknown parameter μ follows therefore as the weighted mean of the information $\hat{\mu}$ coming from the data and of the prior information μ_p. The weight $(V(\mu))^{-1}$ of μ results from the sum of the two weights.

If both the expected value μ and the variance σ^2 of the observations are unknown, the posterior densities of μ and σ^2 are derived in Example 1 of Section 224. Δ

212 Recursive Application

If different and independent vectors $\mathbf{y}_i$ of observations have been collected with $i \in \{1,\ldots,m\}$, Bayes' theorem may be applied recursively. With the first vector $\mathbf{y}_1$ we obtain from (211.1)

$$p(\boldsymbol{\theta}|\mathbf{y}_1) \propto p(\boldsymbol{\theta})p(\mathbf{y}_1|\boldsymbol{\theta}) .$$

Using the posterior density function $p(\boldsymbol{\theta}|\mathbf{y}_1)$ as prior density for the second vector $\mathbf{y}_2$ of observations gives

$$p(\boldsymbol{\theta}|\mathbf{y}_1,\mathbf{y}_2) \propto p(\boldsymbol{\theta}|\mathbf{y}_1)p(\mathbf{y}_2|\boldsymbol{\theta})$$

or with k vectors of observations

$$p(\boldsymbol{\theta}|\mathbf{y}_1,\ldots,\mathbf{y}_k) \propto p(\boldsymbol{\theta}|\mathbf{y}_1,\ldots,\mathbf{y}_{k-1})p(\mathbf{y}_k|\boldsymbol{\theta}) \quad \text{for} \quad k \in \{2,\ldots,m\}, \tag{212.1}$$

where

$$p(\boldsymbol{\theta}|\mathbf{y}_1) \propto p(\boldsymbol{\theta})p(\mathbf{y}_1|\boldsymbol{\theta}) . \tag{212.2}$$

Thus, with each experiment, that is with each observation vector $\mathbf{y}_i$, the knowledge on the unknown parameters $\boldsymbol{\theta}$ is updated, which is equivalent to the process of learning from the gain on experience.

Of course, we do not have to work recursively, but we may analyze all observations at the same time. The likelihood function of the observation vectors $\mathbf{y}_1$ to $\mathbf{y}_k$ is given by $p(\mathbf{y}_1|\boldsymbol{\theta})p(\mathbf{y}_2|\boldsymbol{\theta})\ldots p(\mathbf{y}_k|\boldsymbol{\theta})$ (Koch 1988a, p.107), since the vectors $\mathbf{y}_i$ and $\mathbf{y}_j$ are independent for $i \neq j$. With Bayes' theorem (211.1) we thus obtain

$$p(\boldsymbol{\theta}|\mathbf{y}_1,\ldots,\mathbf{y}_k) \propto p(\boldsymbol{\theta})p(\mathbf{y}_1|\boldsymbol{\theta})p(\mathbf{y}_2|\boldsymbol{\theta})\ldots p(\mathbf{y}_k|\boldsymbol{\theta}) , \tag{212.3}$$

which is identical with (212.1) after substituting (212.2).

An example for the recursive application of Bayes' theorem is given in Section 318.

22 Prior Density Functions

221 Unknown Parameters

The unknown parameters in Bayes' theorem (211.1) are defined as random variables. But this does not mean that the parameters cannot represent constants, like the coordinates of fixed points. The prior and posterior density functions of the unknown parameters describe the state of knowledge on the parameters. By means of these distributions the probability is determined that the parameters cover certain regions. This probability expresses the subjective belief on the unknown parameters, and it needs not be interpreted as a frequency, that means a measurable quantity, since the parameters are not defined by actual or hypothetical random experiments. Only theoretically a probability is assigned to them, to express the state of knowledge. We may therefore interpret the parameters as fixed quantities or variable quantities, variable for instance with time. When applying Bayes' theorem we obtain the posterior density function according to the values the parameters had when the data were taken.

There are situations when no prior information exists on the parameters. Hence, we have to decide how to express ignorance in the prior distribution. This is accomplished by the so-called *noninformative* or *vague prior density functions*, which will be treated next.

222 Noninformative Priors

If we do not know anything about an unknown parameter θ, it may take values from $-\infty$ to $+\infty$. Hence, the appropriate prior probability density function is

$$p(\theta) \propto \text{const.} \quad \text{with} \quad -\infty < \theta < \infty. \tag{222.1}$$

This is an improper density function because $\int_{-\infty}^{\infty} p(\theta)d\theta \neq 1$. But the posterior density function can be normalized, so this is not a serious drawback.

If a parameter assumes values only between 0 and ∞, for example the variance σ^2, we take

$$\theta = \ln\sigma^2, \tag{222.2}$$

where $\ln\sigma^2$ denotes the natural logarithm of σ^2, and again

$$p(\theta) \propto \text{const.} \quad \text{with} \quad -\infty < \theta < \infty.$$

By transforming the variable from θ to σ^2 with (222.2) (Koch 1988a, p.108), we obtain with $d\theta/d\sigma^2=1/\sigma^2$ the noninformative prior density function for the variance σ^2

$$p(\sigma^2) \propto 1/\sigma^2 \quad \text{with} \quad 0 < \sigma^2 < \infty. \tag{222.3}$$

Example 1: The prior density function (222.1) and the meaning of the transformation (222.2), which leads from (222.1) to (222.3), shall be demonstrated by an example. We assume n independent observations y_i, each being normally distributed with the unknown expected value μ and the known variance σ^2. The likelihood function follows from (211.14) by

$$p(\mathbf{y}|\mu) \propto \exp[-\frac{n}{2\sigma^2}(\mu-\hat{\mu})^2].$$

The graph of this likelihood function is a normal curve, whose location on the μ axis depends on $\hat{\mu}$, which means on the data y_i according to (211.11). Different sets of observations change only the location and not the shape of the likelihood curve. The likelihood is *data translated* (Box and Tiao 1973, p.26). If in such a situation no prior information is available for the unknown parameter μ, so that the data have to decide on μ, the appropriate prior density is a constant as formulated by (222.1).

If, on the other hand, the expected value μ is known and the variance σ^2 is unknown, we find with (211.14) the likelihood function

$$p(\mathbf{y}|\sigma^2) \propto (\sigma^2)^{-n/2} \exp[-(n-1)\hat{\sigma}^2/2\sigma^2].$$

The graph of this function changes its shape and location on the σ^2 axis with varying sets of observations through $\hat{\sigma}^2$ according to (211.12). However, by applying the transformation (222.2) we find with $\sigma^2=\exp\theta$ and $d\sigma^2/d\theta=\exp\theta=\sigma^2$ the likelihood function

$$p(\mathbf{y}|\ln\sigma^2) \propto (\sigma^2)^{-(n-2)/2} \exp[-(n-1)\hat{\sigma}^2/2\sigma^2].$$

We may multiply by the constant $(\hat{\sigma}^2)^{(n-2)/2}$ and obtain

$$p(\mathbf{y}|\ln\sigma^2) \propto \exp\{\ln(\sigma^2/\hat{\sigma}^2)^{-(n-2)/2} - \frac{n-1}{2}\exp[\ln(\hat{\sigma}^2/\sigma^2)]\}$$

or finally

$$p(\mathbf{y}|\ln\sigma^2) \propto \exp\{-\frac{n-2}{2}(\ln\sigma^2-\ln\hat{\sigma}^2) - \frac{n-1}{2}\exp[-(\ln\sigma^2-\ln\hat{\sigma}^2)]\}.$$

The graph of this likelihood function only changes its location on the $\ln\sigma^2$ axis, if the data are varied through $\ln\hat{\sigma}^2$. If no prior information is available for $\ln\sigma^2$, the appropriate prior distribution for $\ln\sigma^2$ is therefore a constant or the prior distribution (222.3) for σ^2. Δ

Often it is convenient to parameterize according to a weight or precision parameter τ with

$$\tau = 1/\sigma^2. \tag{222.4}$$

By transforming σ^2 to τ we find with $d\sigma^2/d\tau = -1/\tau^2$ instead of (222.3)

$$p(\tau) \propto 1/\tau \quad \text{with} \quad 0 < \tau < \infty. \tag{222.5}$$

The density (222.3) is invariant to transformations of the form

$$\kappa = (\sigma^2)^n, \tag{222.6}$$

which can already be recognized by (222.5). To show this, we transform σ^2 to κ and find with $\sigma^2 = \kappa^{1/n}$ and $d\sigma^2/d\kappa = n^{-1}\kappa^{1/n-1}$

$$d\sigma^2/\sigma^2 \propto d\kappa/\kappa \tag{222.7}$$

and instead of (222.3)

$$p(\kappa) \propto 1/\kappa \quad \text{with} \quad 0 < \kappa < \infty. \tag{222.8}$$

The differential probability computed with the posterior density $p(\sigma^2|y)$ for σ^2 is obtained with Bayes' theorem (211.1), if (222.3) is used as prior density, by

$$p(\sigma^2|y)d\sigma^2 \propto \frac{1}{\sigma^2}\, p(y|\sigma^2)d\sigma^2.$$

If, on the other hand, the variable κ of (222.6) is chosen, we find with (222.8)

$$p(\kappa|y)d\kappa \propto \frac{1}{\kappa}\, p(y|\sigma^2)d\kappa.$$

However, because of (222.7)

$$p(\sigma^2|y)d\sigma^2 \propto p(\kappa|y)d\kappa, \tag{222.9}$$

so that the posterior distributions give identical probabilities independent of the parameters σ^2 or κ. The prior density (222.3) is therefore invariant to transformations of the form (222.6).

The results (222.1) and (222.3) can be derived from a more general formula which is due to Jeffreys (1961, p.179) and which is also based on the invariance of a transformation. A noninformative prior density function $p(\boldsymbol{\theta})$ of the parameters $\boldsymbol{\theta}$ is given by

$$p(\boldsymbol{\theta}) \propto (\det \mathbf{I}_{\theta})^{1/2}, \tag{222.10}$$

where $\mathbf{I}_{\theta}$ is Fisher's *information matrix* for the parameters $\boldsymbol{\theta}$ associated with the likelihood function. With $\mathbf{I}_{\theta} = (I_{ij})$, $\boldsymbol{\theta} = (\theta_i)$ and $i, j \in \{1, \ldots, u\}$ we have

$$I_{ij} = -E\left(\frac{\partial^2 \ln p(\mathbf{y}|\boldsymbol{\theta})}{\partial\theta_i \partial\theta_j}\right) \quad \text{for} \quad i,j \in \{1,\ldots,u\}, \tag{222.11}$$

where the expectation E is taken with respect to the distribution for **y**.

Example 2: The likelihood function (211.10) of the Example 1 of Section 211 depends on the unknown parameter μ. Taking its natural logarithm gives

$$\ln p(\mathbf{y}|\mu) = -\frac{n}{2}\ln 2\pi - \frac{n}{2}\ln\sigma^2 - \frac{1}{2\sigma^2}\sum_{i=1}^{n}(y_i-\mu)^2.$$

Furthermore,

$$\partial \ln p(\mathbf{y}|\mu)/\partial\mu = \frac{1}{\sigma^2}\sum_{i=1}^{n}(y_i-\mu)$$

and

$$\partial^2 \ln p(\mathbf{y}|\mu)/\partial\mu^2 = -n/\sigma^2$$

and therefore with (222.10)

$$p(\mu) \propto \text{const.}$$

in agreement with (222.1).

We will now assume that the variance σ^2 in (211.10) is the unknown parameter. We parameterize according to (222.4), i.e. $\tau=1/\sigma^2$, and obtain

$$p(\mathbf{y}|\tau) = (2\pi)^{-n/2}\tau^{n/2}\exp[-\frac{\tau}{2}\sum_{i=1}^{n}(y_i-\mu)^2].$$

Taking the natural logarithm gives

$$\ln p(\mathbf{y}|\tau) = -\frac{n}{2}\ln 2\pi + \frac{n}{2}\ln\tau - \frac{\tau}{2}\sum_{i=1}^{n}(y_i-\mu)^2$$

and

$$\partial \ln p(\mathbf{y}|\tau)/\partial\tau = \frac{n}{2\tau} - \frac{1}{2}\sum_{i=1}^{n}(y_i-\mu)^2.$$

Furthermore, we have

$$\partial^2 \ln p(\mathbf{y}|\tau)/\partial\tau^2 = -\frac{n}{2\tau^2}$$

and with (222.10)

$$p(\tau) \propto 1/\tau$$

in agreement with (222.5).

If in (211.10) both μ and τ are considered as unknown, we obtain from (222.11) with $E(y_i)=\mu$

$$I_{11} = n\tau, \quad I_{12} = -E(\sum_{i=1}^{n}(y_i-\mu)) = 0,$$

$$I_{21} = -E(\sum_{i=1}^{n}(y_i-\mu)) = 0, \quad I_{22} = n/(2\tau^2)$$

and from (222.10)

$$p(\mu,\tau) \propto 1/\sqrt{\tau},$$

which does not agree with (222.5). Thus, in order to obtain coinciding results, the unknown parameters μ and σ^2 have to be considered as being independent. Δ

The invariance property of Jeffreys′ prior density function (222.10) can be shown by means of the

Theorem: Let $\boldsymbol{\beta}=\mathbf{t}(\boldsymbol{\theta})$ with $\mathbf{t}(\boldsymbol{\theta})=(t_i(\boldsymbol{\theta}))$ and $i\in\{1,\ldots,u\}$ be an injective differentiable transformation and $\mathbf{I}_\beta$ and $\mathbf{I}_\theta$ the information matrices for $\boldsymbol{\beta}$ and $\boldsymbol{\theta}$, then

$$(\det\mathbf{I}_\beta)^{1/2}d\boldsymbol{\beta} = (\det\mathbf{I}_\theta)^{1/2}d\boldsymbol{\theta}. \tag{222.12}$$

Proof (Zellner 1971, p.48): We will first show by abbreviating $p(\mathbf{y}|\boldsymbol{\theta})$ with p the relation

$$-E(\frac{\partial^2 \ln p}{\partial\theta_i\partial\theta_j}) = E(\frac{\partial \ln p}{\partial\theta_i}\frac{\partial \ln p}{\partial\theta_j}) \quad \text{for} \quad i,j\in\{1,\ldots,u\}. \tag{222.13}$$

By differentiation we find

$$-E(\frac{\partial^2 \ln p}{\partial\theta_i\partial\theta_j}) = -E(\partial(\frac{1}{p}\frac{\partial p}{\partial\theta_i})/\partial\theta_j) = -E(-\frac{1}{p^2}\frac{\partial p}{\partial\theta_i}\frac{\partial p}{\partial\theta_j} + \frac{1}{p}\frac{\partial^2 p}{\partial\theta_i\partial\theta_j})$$

$$= E(\frac{\partial \ln p}{\partial\theta_i}\frac{\partial \ln p}{\partial\theta_j}) - \int_{-\infty}^{\infty}\ldots\int_{-\infty}^{\infty}\frac{p}{p}\frac{\partial^2 p}{\partial\theta_i\partial\theta_j}d\mathbf{y}$$

$$= E(\frac{\partial \ln p}{\partial\theta_i}\frac{\partial \ln p}{\partial\theta_j}) - \frac{\partial^2}{\partial\theta_i\partial\theta_j}\int_{-\infty}^{\infty}\ldots\int_{-\infty}^{\infty}p\,d\mathbf{y}$$

$$= E(\frac{\partial \ln p}{\partial\theta_i}\frac{\partial \ln p}{\partial\theta_j}),$$

since the value of the integral on the right-hand side is equal to one (Koch 1988a, p.104).

By differentiating $\ln p(\mathbf{y}|\boldsymbol{\theta})$ with respect to the new variable β_i with $\boldsymbol{\beta}=(\beta_i)$ we obtain

$$\frac{\partial \ln p}{\partial\beta_i} = \sum_{k=1}^{u}\frac{\partial \ln p}{\partial\theta_k}\frac{\partial\theta_k}{\partial\beta_i}$$

and

$$\frac{\partial \ln p}{\partial \beta_i}\frac{\partial \ln p}{\partial \beta_j} = \sum_{k=1}^{u}\sum_{l=1}^{u}\frac{\partial \theta_k}{\partial \beta_i}\frac{\partial \ln p}{\partial \theta_k}\frac{\partial \ln p}{\partial \theta_l}\frac{\partial \theta_l}{\partial \beta_j} .$$

By defining

$$I_{\beta,i,j} = E(\frac{\partial \ln p}{\partial \beta_i}\frac{\partial \ln p}{\partial \beta_j}) \quad \text{and} \quad I_{\theta,k,l} = E(\frac{\partial \ln p}{\partial \theta_k}\frac{\partial \ln p}{\partial \theta_l})$$

we obtain

$$I_{\beta,i,j} = \sum_{k=1}^{u}\sum_{l=1}^{u}\frac{\partial \theta_k}{\partial \beta_i} I_{\theta,k,l}\frac{\partial \theta_l}{\partial \beta_j} .$$

With (222.11) and (222.13) we recognize

$$\mathbf{I}_\beta = (I_{\beta,i,j}) \text{ and } \mathbf{I}_\theta = (I_{\theta,k,l}).$$

Thus, by defining $\mathbf{J}=(\partial\theta_o/\partial\beta_p)$ with $o,p\in\{1,\ldots,u\}$ we obtain

$$\mathbf{I}_\beta = \mathbf{J}'\mathbf{I}_\theta\mathbf{J}$$

and

$$(\det\mathbf{I}_\beta)^{1/2} = (\det\mathbf{I}_\theta)^{1/2}\det\mathbf{J}.$$

Furthermore, we have (Koch 1988a, p.85)

$$d\boldsymbol{\theta} = |\det\mathbf{J}|d\boldsymbol{\beta},$$

since $\det\mathbf{J}$ is the Jacobian of the transformation which is inverse to $\boldsymbol{\beta}=\mathbf{t}(\boldsymbol{\theta})$. Hence,

$$(\det\mathbf{I}_\beta)^{1/2}d\boldsymbol{\beta} = (\det\mathbf{I}_\theta)^{1/2}d\boldsymbol{\theta},$$

which proves the theorem. □

The invariance of the prior density function (222.10) is equivalent to the invariance demonstrated in (222.9). This can be shown by using (222.10) as prior density in Bayes' theorem (211.1). We obtain with the posterior density $p(\boldsymbol{\theta}|\mathbf{y})$ of the parameters $\boldsymbol{\theta}$ the differential probability

$$p(\boldsymbol{\theta}|\mathbf{y})d\boldsymbol{\theta} \propto (\det\mathbf{I}_\theta)^{1/2}p(\mathbf{y}|\boldsymbol{\theta})d\boldsymbol{\theta}.$$

If the parameters $\boldsymbol{\beta}$ of (222.12) are chosen, we find

$$p(\boldsymbol{\beta}|\mathbf{y})d\boldsymbol{\beta} \propto (\det\mathbf{I}_\beta)^{1/2}p(\mathbf{y}|\boldsymbol{\theta})d\boldsymbol{\beta}.$$

Because of (222.12) we have

$$p(\boldsymbol{\theta}|\mathbf{y})d\boldsymbol{\theta} \propto p(\boldsymbol{\beta}|\mathbf{y})d\boldsymbol{\beta}, \tag{222.14}$$

so that the posterior distributions give identical probabilities independent of the parameters $\boldsymbol{\theta}$ or $\boldsymbol{\beta}$. The prior density (222.10) is therefore invariant to transformations $\boldsymbol{\beta}=\mathbf{t}(\boldsymbol{\theta})$.

Example 3: We go back to the Example 1 of Section 211 and introduce instead of an informative prior for the unknown parameter μ the noninformative prior

$$p(\mu) \propto \text{const.}$$

from (222.1). Together with the likelihood function (211.14) we obtain from Bayes' theorem (211.7) after neglecting the constants the posterior density for μ

$$p(\mu|\mathbf{y}) \propto \exp[-\frac{n}{2\sigma^2}(\mu-\hat{\mu})^2]. \tag{222.15}$$

It becomes obvious with (A11.1) that this posterior density function is normal with the expected value $\hat{\mu}$ and the variance σ^2/n, hence

$$E(\mu) = \hat{\mu} \quad \text{and} \quad V(\mu) = \sigma^2/n. \tag{222.16}$$

The same result is obtained with $\sigma^2_\mu \to \infty$ in (211.18) and (211.19). Geometrically it means that the prior density function (211.8) of Example 1 of Section 211 is spread out by $\sigma^2_\mu \to \infty$, leading in the limit to a noninformative prior. Δ

223 Maximum Entropy Priors

Entropy is a measure of uncertainty. By applying the principle of maximum entropy we may select distributions which contain the largest amount of uncertainty compatible with given constraints. Since prior information is generally incomplete, beyond the given information the prior information should be as uncertain or noninformative as possible. If, for instance, the expected value and the variance of a random variable are given, among all possible distributions we want to select the one which contains the largest amount of uncertainty compatible with the known expected value and the known variance.

We are looking for a real-valued function I, which measures the *information* gained by the random event A of an experiment. The information which is gained can be interpreted as the *uncertainty* which is removed and which was existing before the outcome of the experiment. Hence, the uncertainty of the outcome of an experiment resulting in an event A is equal to the information gained, if the experiment leads to the event A.

The function $I(A)$ measuring the information or uncertainty of an event A should satisfy the following properties:

1. If an event A has probability one of occurring, $P(A)=1$, then its uncertainty is zero, $I(A)=0$.
2. If $P(A_1)<P(A_2)$, then $I(A_1)>I(A_2)$, that is, the first event has larger uncertainty.
3. If A_1 and A_2 are independent, then $I(A_1 \cap A_2)=I(A_1)+I(A_2)$.

The measure I has to depend on the probability of events. Hence, we are looking for a function, say $G(P(A))$, defined on the interval $(0,1)$. If it is a monotonic decreasing function on $(0,1)$, vanishing at 1, it fulfills 1. and 2. If A_1 and A_2 are independent, then $P(A_1 \cap A_2)=P(A_1)P(A_2)$ and with 3. we obtain

$$G(P(A_1)P(A_2)) = G(P(A_1)) + G(P(A_2)). \tag{223.1}$$

The only function satisfying (223.1) is the natural logarithm multiplied by a constant, hence $G(P(A))=c\ln P(A)$. This function vanishes at 1, but is increasing, so that we have to choose a negative constant. With $I(A)=G(P(A))$ the measure $I(A)$ of information or uncertainty of an event A therefore follows by

$$I(A) = - c \ln P(A). \tag{223.2}$$

Now we interpret $P(A)$ as the probability density $p(x_i|\theta)$ of a discrete random variable x with values x_i defined in the probability space of an experiment. Thus,

$$p(x_i|\theta) \geq 0 \quad \text{for} \quad i \in \{1,\dots,n\} \quad \text{and} \quad \sum_{i=1}^{n} p(x_i|\theta) = 1,$$

where θ is the parameter of the density function. The expected value of the information or uncertainty of the experiment is obtained with $c=1$ in (223.2) by

$$H_n = - \sum_{i=1}^{n} p(x_i|\theta) \ln p(x_i|\theta). \tag{223.3}$$

The expected value H_n is called *discrete entropy*.

We assume $p(x_i|\theta) \ln p(x_i|\theta)=0$ for $p(x_i|\theta)=0$ according to the limit $\lim_{x \to 0} x \ln x=0$.

Let a continuous random variable x be defined by an experiment on the interval $[a,b]$ with the probability density function $p(x|\theta)$ depending on θ, thus

$$p(x|\theta) \geq 0 \quad \text{and} \quad \int_a^b p(x|\theta)dx = 1.$$

The *continuous entropy* is defined correspondingly to (223.3) by

$$H = - \int_a^b p(x|\theta) \ln p(x|\theta)dx. \tag{223.4}$$

The properties of the discrete and the continuous entropy are different, for instance, while $H_n \geq 0$, H may be also negative. This can be shown by assuming the uniform distribution

$$p(x|a,b) = \frac{1}{b-a} \quad \text{for} \quad a \leq x \leq b,$$

which gives

$$H = \frac{b-a}{b-a} \ln(b-a)$$

with

$$\ln(b-a) < 0 \quad \text{for} \quad (b-a) < 1.$$

The interpretation of the discrete and continuous entropy as measures of uncertainty is therefore different. However, a concurring interpretation in the discrete and continuous case is obtained, if the entropy is interpreted as a variation of information, if one passes from an initial probability measure given by the uniform distribution on the interval $[a,b]$ to a new probability measure defined by the probability density function $p(x_i|\theta)$ and $p(x|\theta)$, respectively (Guiasu 1977, p.28).

We use now the principle of maximum entropy to derive distributions, that is we maximize the entropy subject to some constraints. This will lead us to distributions which contain maximum uncertainty compatible with the constraints. These distributions are therefore well suited as prior distributions. The following theorem summarizes the results.

Theorem: The density function of a random variable x, which is defined on the interval $[a,b]$ and which maximizes the entropy, is the density of the uniform distribution

$$p(x|a,b) = 1/(b-a) \quad \text{for} \quad a \leq x \leq b. \tag{223.5}$$

The density function of a random variable x with the expected value $E(x)=\mu$ and the variance $V(x)=\sigma^2$, which is defined on the interval $(-\infty,\infty)$ and which maximizes the entropy, is the density of the normal distribution

$$p(x|\mu,\sigma^2) = \frac{1}{\sqrt{2\pi}\,\sigma} e^{-(x-\mu)^2/2\sigma^2} \quad \text{for} \quad -\infty < x < \infty. \tag{223.6}$$

The density function of a random variable x with the expected value $E(x)=\mu$, which is defined on the interval $[0,\infty)$ and which maximizes the entropy, is the density of the exponential distribution

$$p(x|\mu) = \frac{1}{\mu} e^{-x/\mu} \quad \text{for} \quad 0 \leq x < \infty. \tag{223.7}$$

The density function of a random variable x with the expected value $E(x)=\mu$ and the variance $V(x)=\sigma^2$, which is defined on the interval $[0,\infty)$ and which maximizes the entropy, is the density of the truncated normal distribution

$$p(x|\mu,\sigma^2) = \exp(-k_o)\exp[-k_1x-k_2(x-\mu)^2] \text{ for } k_2 > 0,\ \sigma^2 < \mu^2$$

$$\text{and } 0 \le x < \infty \tag{223.8}$$

with

$$\exp(k_o) = \int_0^\infty \exp[-k_1x-k_2(x-\mu)^2]dx$$

$$\exp(-k_o)\int_0^\infty x\exp[-k_1x-k_2(x-\mu)^2]dx = \mu$$

$$\exp(-k_o)\int_0^\infty (x-\mu)^2\exp[-k_1x-k_2(x-\mu)^2]dx = \sigma^2.$$

Proof: We will first derive a general solution which contains the given distributions as special cases. Thus, the entropy H in (223.4)

$$H = -\int_a^b p(x|\theta)\ln p(x|\theta)dx \tag{223.9}$$

for the distribution of a random variable x defined on the interval $[a,b]$ has to be maximized subject to the constraint of the normalization in (211.6)

$$\int_a^b p(x|\theta)dx = 1 \tag{223.10}$$

and subject to the constraints

$$\int_a^b f_i(x)p(x|\theta)dx = \mu_i \quad \text{for} \quad i\in\{1,\dots,n\} \tag{223.11}$$

resulting from given expected values μ_i, like the mean or the variance, for some functions $f_i(x)$.

For computing the extreme value of the entropy we introduce the Lagrange function (Koch 1988a, p.80)

$$w(x) = -\int_a^b p(x|\theta)\ln p(x|\theta)dx - k_o[\int_a^b p(x|\theta)dx-1]$$

$$-\sum_{i=1}^n k_i[\int_a^b f_i(x)p(x|\theta)dx-\mu_i],$$

where $-k_o$ and $-k_i$ denote the Lagrange multipliers. Since we look for the maximum of $w(x)$, we may neglect constant terms which change the height of the maximum but not its position. Hence,

$$\int_a^b p(x|\theta)[-\ln p(x|\theta) - k_o - \sum_{i=1}^n k_if_i(x)]dx$$

$$= \int_a^b p(x|\theta) \ln\{\frac{1}{p(x|\theta)} \exp[-k_o - \sum_{i=1}^n k_i f_i(x)]\}dx$$

$$\le \int_a^b p(x|\theta)\{\frac{1}{p(x|\theta)} \exp[-k_o - \sum_{i=1}^n k_i f_i(x)] - 1\}dx,$$

where the inequality holds because of

$$\ln x = x - 1 \quad \text{for} \quad x = 1$$

and

$$\ln x < x - 1 \quad \text{for} \quad x > 0 \quad \text{and} \quad x \neq 1.$$

The equality follows with

$$p(x|\theta) = \exp[-k_o - \sum_{i=1}^n k_i f_i(x)]. \qquad (223.12)$$

For this case the right-hand side of the inequality is constant, so that an upper limit, that is the maximum, of the Lagrange function is found. By substituting the result (223.12) in (223.10) the constant k_o is determined. We have

$$\int_a^b \exp[-k_o - \sum_{i=1}^n k_i f_i(x)]dx = 1$$

and therefore

$$\exp(k_o) = \int_a^b \exp[-\sum_{i=1}^n k_i f_i(x)]dx. \qquad (223.13)$$

With this result the density function is obtained by

$$p(x|\theta) = \exp(-k_o) \exp[-\sum_{i=1}^n k_i f_i(x)]. \qquad (223.14)$$

We will now assume a random variable x defined on the interval $[a,b]$ subject to no constraints (223.11), so that $k_i=0$ in (223.13) and (223.14). We obtain from (223.14)

$$p(x|\theta) = \exp(-k_o)$$

and from (223.13)

$$\exp(k_o) = \int_a^b dx = b - a$$

or

$$k_o = \ln(b-a).$$

Substituting this result in the density function leads to

$$p(x|a,b) = e^{-\ln(b-a)} = 1/(b-a),$$

which proves (223.5).

Now we assume a random variable x with the expected value $E(x)=\mu$ and the variance $V(x)=\sigma^2$ defined on the interval $(-\infty,\infty)$. We therefore have

$$f_1(x) = x,\ \mu_1 = \mu,\ f_2(x) = (x-\mu)^2,\ \mu_2 = \sigma^2$$

in (223.11) and the first constraint gives

$$\int_{-\infty}^{\infty} xp(x|\theta)dx = \mu.$$

We change the variable from x to y with

$$x = y + \mu \quad \text{and} \quad dx/dy = 1$$

and obtain instead

$$\int_{-\infty}^{\infty} yp(y|\theta)dy = 0.$$

Hence, we find $k_1=0$ in (223.13) and (223.14) for the variable y and obtain the density function

$$p(y|\theta) = \exp(-k_o)\ \exp(-k_2y^2)$$

with

$$\exp(k_o) = \int_{-\infty}^{\infty} \exp(-k_2y^2)dy.$$

The integration gives (Gradshteyn and Ryzhik 1965, p.307)

$$k_o = \ln \int_{-\infty}^{\infty} \exp(-k_2y^2)dy = \ln(\pi/k_2)^{1/2}$$

and therefore

$$p(y|\theta) = (k_2/\pi)^{1/2}\exp(-k_2y^2).$$

The second constraint of (223.11) for y gives (Gradshteyn and Ryzhik 1965, p.337)

$$(k_2/\pi)^{1/2} \int_{-\infty}^{\infty} y^2 \exp(-k_2y^2)dy = (k_2/\pi)^{1/2}(1/2k_2)(\pi/k_2)^{1/2} = \sigma^2$$

or

$$k_2 = 1/2\sigma^2.$$

Thus, changing back to x with $y=x-\mu$ we obtain

$$p(x|\mu,\sigma^2) = \frac{1}{\sqrt{2\pi}\,\sigma}\ e^{-(x-\mu)^2/2\sigma^2},$$

which proves (223.6) because of (A11.1).

We now introduce a random variable x with the expected value $E(x)=\mu$ defined on the interval $[0,\infty)$. We therefore have

$$f_1(x) = x \quad \text{and} \quad \mu_1 = \mu$$

in (223.11) and find from (223.14) the density function

$$p(x|\theta) = \exp(-k_o)\,\exp(-k_1 x) \qquad (223.15)$$

with

$$\exp(k_o) = \int_0^\infty \exp(-k_1 x)dx$$

from (223.13). The integration leads to

$$\exp(k_o) = [-\frac{1}{k_1}\exp(-k_1 x)]_0^\infty = \frac{1}{k_1}$$

so that

$$p(x|\theta) = k_1\,\exp(-k_1 x).$$

The constant k_1 follows from the first constraint of (223.11)

$$k_1 \int_0^\infty x\,\exp(-k_1 x)dx = \mu.$$

The integration gives

$$k_1[\frac{\exp(-k_1 x)}{k_1^2}(-k_1 x-1)]_0^\infty = k_1/k_1^2 = \mu$$

or

$$k_1 = 1/\mu.$$

We therefore find the exponential distribution

$$p(x|\mu) = \frac{1}{\mu}\,e^{-x/\mu},$$

which proves (223.7). As shown, if x has the exponential distribution (223.7), its expected value $E(x)$ is given by $E(x)=\mu$. The variance follows with (A12.5) from

$$V(x) = \frac{1}{\mu}\int_0^\infty x^2 e^{-x/\mu}dx-\mu^2 = \frac{1}{\mu}\,[e^{-x/\mu}(-\mu x^2-2\mu^2 x-2\mu^3)]_0^\infty - \mu^2 = \mu^2. \qquad (223.16)$$

We finally assume a random variable x with the expected value $E(x)=\mu$ and the variance $V(x)=\sigma^2$ defined on the interval $[0,\infty)$. Thus, we have

$$f_1(x) = x,\ \mu_1 = \mu,\ f_2(x) = (x-\mu)^2,\ \mu_2 = \sigma^2$$

in (223.11) and (223.14) gives the density

$$p(x|\mu,\sigma^2) = \exp(-k_o)\ \exp[-k_1x-k_2(x-\mu)^2].$$

We require $p(x|\mu,\sigma^2)$ to be a proper density, so that we must have $k_2>0$, because otherwise the integral over the density function takes on an infinite value. Transforming the exponent gives

$$p(x|\mu,\sigma^2) = \exp(-k_o)\ \exp[-k_2(\frac{k_1}{k_2}\ x+x^2-2\mu x+\mu^2)]$$

$$= \exp(-k_o)\ \exp[-k_2(x^2-2(\mu - \frac{k_1}{2k_2})x+\mu^2)]$$

$$= \exp(-k_o)\ \exp[-k_2(\mu^2-(\mu - \frac{k_1}{2k_2})^2)]\exp[-k_2(x-(\mu - \frac{k_1}{2k_2}))^2]. \qquad (223.17)$$

Except for the normalization constant the density $p(x|\mu,\sigma^2)$ has the form of a normal distribution with the expected value $\mu-k_1/2k_2$ and the variance $1/2k_2$, since $k_2>0$. The density therefore stems from a truncated normal distribution defined on the interval $[0,\infty)$. It can be shown (Dowson and Wragg 1973) that the density function exists for $\sigma^2<\mu^2$. For this case the normalization constant follows from (223.13) by

$$\exp(k_o) = \int_0^\infty \exp[-k_1x-k_2(x-\mu)^2]dx.$$

The constants k_1 and k_2 are determined with (223.11) by

$$\exp(-k_o) \int_0^\infty x\ \exp[-k_1x-k_2(x-\mu)^2]dx = \mu$$

and

$$\exp(-k_o) \int_0^\infty (x-\mu)^2\ \exp[-k_1x-k_2(x-\mu)^2]dx = \sigma^2,$$

so that (223.8) is proved.

It should be noted that for $k_2=0$ we obtain (223.15) and thus the exponential distribution, for which according to (223.16) $\sigma^2=\mu^2$ is valid. If σ^2 approaches μ^2, the truncated normal distribution therefore adopts the shape of the exponential distribution. If, on the other hand, σ^2 goes to zero, the probability mass of the truncated normal distribution concentrates around the expected value and the truncated part becomes meaningless. The truncated normal distribution then adopts the shape of the normal distribution. □

The theorem (223.5) confirms the noninformative prior density function (222.1), which introduces a constant for the density function of a parameter for which no prior information is available.

The theorem (223.6) emphasizes the importance of the normal distribution as a prior density. According to (A11.3) the density function of a normally distributed random variable is uniquely determined by its expected value and its variance. It is therefore a remarkable fact that if the expected value and the variance of a random variable are given, the normal distribution contains among all distributions the largest amount of uncertainty which is compatible with the given two values. The normal distribution should therefore be introduced as a prior distribution for a parameter which is defined for the whole real axis and for which prior information is available on its expected value and variance.

If a parameter like a variance is defined for positive values only and if its expected value and its variance are known in advance, then the truncated normal distribution (223.8) should be taken as a prior distribution provided $\sigma^2<\mu^2$. This distribution degenerates to an exponential distribution in the case of $\sigma^2=\mu^2$, as shown above. The constants k_1 and k_2 of the truncated normal distribution have to be determined numerically, for instance by numerical iterations. For this procedure the integrals of (223.8) defining $\exp(k_0)$, μ and σ^2 need to be computed. By taking the representation (223.17) for the density of the truncated normal distribution the following integrals have to be computed

$$I_1 = \int_0^\infty \exp[-k_2(x-\mu+k_1/2k_2)^2]dx,$$

$$I_2 = \int_0^\infty x \exp[-k_2(x-\mu+k_1/2k_2)^2]dx,$$

$$I_3 = \int_0^\infty (x-\mu)^2 \exp[-k_2(x-\mu+k_1/2k_2)^2]dx. \qquad (223.18)$$

To solve the first integral we substitute

$$y = \sqrt{k_2}\,(x-\mu+k_1/2k_2) \quad \text{with} \quad dy = \sqrt{k_2}\,dx$$

and find

$$I_1 = \frac{1}{\sqrt{k_2}} \int_{\sqrt{k_2}(-\mu+k_1/2k_2)}^{\infty} \exp(-y^2)dy$$

$$= \frac{1}{2}\left(\frac{\pi}{k_2}\right)^{1/2} \begin{cases} \{1-\mathrm{erf}[\sqrt{k_2}\,(-\mu+k_1/2k_2)]\}, & \text{if } -\mu + k_1/2k_2 \geq 0 \\ \{1+\mathrm{erf}[\sqrt{k_2}\,(\mu-k_1/2k_2)]\}, & \text{if } -\mu + k_1/2k_2 < 0, \end{cases} \qquad (223.19)$$

where erf denotes the error function (Abramowitz and Stegun 1965, p.297), which can be readily computed, for instance by rational approximations (Abramowitz and Stegun 1965, p.299). However, the absolute value of the argument of the error function should be small. This is no longer ensured, if the truncated normal distribution approaches the normal distribution or the exponential distribution, as explained above. A numerical integration should then be applied.

To compute the second integral I_2 in (223.18) we introduce the identity

$$I_2 = \int_0^\infty (x-\mu+k_1/2k_2)\exp[-k_2(x-\mu+k_1/2k_2)^2]dx + (\mu-k_1/2k_2)I_1 .$$

By a change of variables

$$y = (x-\mu+k_1/2k_2)^2 \quad \text{with} \quad dy = 2(x-\mu+k_1/2k_2)dx$$

we find

$$I_2 = \frac{1}{2} \int_{(-\mu+k_1/2k_2)^2}^\infty \exp(-k_2y)dy + (\mu-k_1/2k_2)I_1$$

$$= \frac{1}{2} \left[-\frac{1}{k_2} \exp(-k_2y) \right]_{(-\mu+k_1/2k_2)^2}^\infty + (\mu-k_1/2k_2)I_1$$

$$= \frac{1}{2k_2} \exp[-k_2(-\mu+k_1/2k_2)^2] + (\mu-k_1/2k_2)I_1 . \tag{223.20}$$

Finally the third integral I_3 of (223.18) gives

$$I_3 = J_3 - 2\mu I_2 + \mu^2 I_1 \tag{223.21}$$

with

$$J_3 = \int_0^\infty x^2 \exp[-k_2(x-\mu+k_1/2k_2)^2]dx .$$

To solve J_3 we integrate I_1 by parts

$$I_1 = \int_0^\infty \exp[-k_2(x-\mu+k_1/2k_2)^2]dx = [x \exp[-k_2(x-\mu+k_1/2k_2)^2]]_0^\infty$$

$$- \int_0^\infty x \exp[-k_2(x-\mu+k_1/2k_2)^2][-2k_2(x-\mu+k_1/2k_2)]dx$$

$$= 2k_2J_3 + 2k_2(-\mu+k_1/2k_2)I_2 .$$

Thus, we obtain

$$J_3 = \frac{1}{2k_2} I_1 - (-\mu+k_1/2k_2)I_2 . \tag{223.22}$$

224 Conjugate Priors

Prior information may also be introduced by *conjugate priors*, also called *natural conjugate priors*. If a conjugate prior belongs to a certain family of distributions, then for any number of observations the posterior density belongs to the same family of distributions. This is the most important property of conjugate priors. As a consequence we are able to work with analytically tractable distributions, since the start with a tractable prior will lead to a tractable posterior distribution.

Only a short explanation of conjugate priors will be given. A more rigorous presentation can be found in DeGroot (1970, p.159), Raiffa and Schlaifer (1961, p.43), Pilz (1983, p.28). Let $t_i = t_i(y)$ with $t = (t_i)$ and $i \in \{1, \ldots, r\}$ be *sufficient statistics*. They condense the information on the unknown parameters $\boldsymbol{\theta}$, which is contained in the data $\mathbf{y}$, into single variables such that no information on $\boldsymbol{\theta}$ is lost. The statistics $\mathbf{t}$ are said to be sufficient for $\boldsymbol{\theta}$, if the likelihood function $p(\mathbf{y}|\boldsymbol{\theta})$ can be factored as

$$p(\mathbf{y}|\boldsymbol{\theta}) = g(\mathbf{t}|\boldsymbol{\theta})h(\mathbf{y}), \tag{224.1}$$

where the function $g(\mathbf{t}|\boldsymbol{\theta})$ is non negative and depends on $\mathbf{y}$ only through $\mathbf{t}$ and $h(\mathbf{y})$ is non negative and does not contain $\boldsymbol{\theta}$. The relation (224.1) is called the factorization theorem (Raiffa and Schlaifer 1961, p.33; Rao 1973, p.131).

A natural conjugate prior $p(\boldsymbol{\theta}|\mathbf{t})$ of the unknown parameters $\boldsymbol{\theta}$ as a function of $\mathbf{t}$ is derived by

$$p(\boldsymbol{\theta}|\mathbf{t}) \propto g(\mathbf{t}|\boldsymbol{\theta}), \tag{224.2}$$

where the density $p(\boldsymbol{\theta}|\mathbf{t})$ obtains its functional form from $g(\mathbf{t}|\boldsymbol{\theta})$, but the roles of the statistics $\mathbf{t}$ and the parameters $\boldsymbol{\theta}$ are interchanged. In $g(\mathbf{t}|\boldsymbol{\theta})$ the parameters $\boldsymbol{\theta}$ are fixed and the statistics $\mathbf{t}$ are random, while in $p(\boldsymbol{\theta}|\mathbf{t})$ the parameters $\boldsymbol{\theta}$ are random and the statistics $\mathbf{t}$ are fixed.

Without mentioning it, a conjugate prior has been already applied in Example 1 of Section 211 for the unknown expected value of independent, identically and normally distributed observations. This can be seen with the likelihood function (211.14), which can be factored according to (224.1) with $h(\mathbf{y})=1$. Using (224.2) with $\mu_p = \hat{\mu}$ and $\sigma_\mu^2 = \sigma^2/n$ leads to the conjugate prior (211.8), which is normal. The posterior distribution (211.17) is also normal. Conjugate priors for the parameters of various standard distributions are derived in DeGroot (1970, p.164), Raiffa and Schlaifer (1961, p.261).

Since we will discuss linear models in Section 31, the conjugate priors for the unknown parameters of these models will be derived here. According to (311.6) the likelihood function follows with

$$p(\mathbf{y}|\boldsymbol{\beta},\sigma^2) = (2\pi\sigma^2)^{-n/2} \exp[-\frac{1}{2\sigma^2}(\mathbf{y}-\mathbf{X}\boldsymbol{\beta})'(\mathbf{y}-\mathbf{X}\boldsymbol{\beta})], \tag{224.3}$$

where $\mathbf{y}$ is the n×1 random vector of observations, $\mathbf{X}$ the n×u matrix of known coefficients with rank$\mathbf{X}$=u, $\boldsymbol{\beta}$ the u×1 vector of unknown parameters and σ^2 the unknown variance factor with $\sigma^2>0$. The statistics or the estimates $\hat{\boldsymbol{\beta}}$ and $\hat{\sigma}^2$ of the unknown parameters $\boldsymbol{\beta}$ and σ^2 are introduced by (Koch 1988a, p.187,192)

$$\hat{\boldsymbol{\beta}} = (\mathbf{X}'\mathbf{X})^{-1}\mathbf{X}'\mathbf{y} \quad \text{and} \quad \hat{\sigma}^2 = \frac{1}{n-u}(\mathbf{y}-\mathbf{X}\hat{\boldsymbol{\beta}})'(\mathbf{y}-\mathbf{X}\hat{\boldsymbol{\beta}}). \tag{224.4}$$

By completing the square of the exponent we find

$$\begin{aligned}(\mathbf{y}-\mathbf{X}\boldsymbol{\beta})'(\mathbf{y}-\mathbf{X}\boldsymbol{\beta}) &= (\mathbf{y}-\mathbf{X}\hat{\boldsymbol{\beta}}-\mathbf{X}(\boldsymbol{\beta}-\hat{\boldsymbol{\beta}}))'(\mathbf{y}-\mathbf{X}\hat{\boldsymbol{\beta}}-\mathbf{X}(\boldsymbol{\beta}-\hat{\boldsymbol{\beta}}))\\ &= (\mathbf{y}-\mathbf{X}\hat{\boldsymbol{\beta}})'(\mathbf{y}-\mathbf{X}\hat{\boldsymbol{\beta}}) + (\boldsymbol{\beta}-\hat{\boldsymbol{\beta}})'\mathbf{X}'\mathbf{X}(\boldsymbol{\beta}-\hat{\boldsymbol{\beta}})\end{aligned} \tag{224.5}$$

because of

$$(\boldsymbol{\beta}-\hat{\boldsymbol{\beta}})'\mathbf{X}'(\mathbf{y}-\mathbf{X}\hat{\boldsymbol{\beta}}) = (\boldsymbol{\beta}-\hat{\boldsymbol{\beta}})'\mathbf{X}'(\mathbf{y}-\mathbf{X}(\mathbf{X}'\mathbf{X})^{-1}\mathbf{X}'\mathbf{y}) = 0.$$

This gives the likelihood function in the form of

$$p(\mathbf{y}|\boldsymbol{\beta},\sigma^2) = (2\pi\sigma^2)^{-n/2}\exp\{-\frac{1}{2\sigma^2}[(n-u)\hat{\sigma}^2 + (\boldsymbol{\beta}-\hat{\boldsymbol{\beta}})'\mathbf{X}'\mathbf{X}(\boldsymbol{\beta}-\hat{\boldsymbol{\beta}})]\}.$$

With $h(\mathbf{y})=1$ we see by the factorization theorem (224.1) that $\hat{\boldsymbol{\beta}}$ and $\hat{\sigma}$ are sufficient statistics for $\boldsymbol{\beta}$ and σ^2. Instead of σ^2 we will use the weight or the precision parameter τ with

$$\tau = 1/\sigma^2 \quad \text{and} \quad \tau > 0, \tag{224.6}$$

so that the likelihood function follows with

$$p(\mathbf{y}|\boldsymbol{\beta},\tau) = (2\pi)^{-n/2}\tau^{n/2} \exp\{-\frac{\tau}{2}[(n-u)\hat{\sigma}^2 + (\boldsymbol{\beta}-\hat{\boldsymbol{\beta}})'\mathbf{X}'\mathbf{X}(\boldsymbol{\beta}-\hat{\boldsymbol{\beta}})]\}. \tag{224.7}$$

The conjugate prior for $\boldsymbol{\beta}$ and τ is now derived by rewriting (224.7) according to (224.2). With

$$\boldsymbol{\mu} = \hat{\boldsymbol{\beta}},\ \mathbf{V}^{-1} = \mathbf{X}'\mathbf{X},\ b = \frac{1}{2}(n-u)\hat{\sigma}^2,\ \frac{u}{2} + p - 1 = \frac{n}{2}$$

and neglecting the constants we obtain the conjugate prior

$$p(\boldsymbol{\beta},\tau|\boldsymbol{\mu},\mathbf{V},b,p) \propto \tau^{u/2+p-1} \exp\{-\frac{\tau}{2}[2b+(\boldsymbol{\beta}-\boldsymbol{\mu})'\mathbf{V}^{-1}(\boldsymbol{\beta}-\boldsymbol{\mu})]\}. \tag{224.8}$$

This is, after introducing the appropriate constants from (A23.1), the density of the normal-gamma distribution

$$\boldsymbol{\beta},\tau \sim NG(\boldsymbol{\mu},\mathbf{V},b,p). \tag{224.9}$$

Now we have to show that using (224.8) as a prior distribution leads to a normal-gamma distribution for the posterior distribution, if the likelihood function of the observations stems from a normal distribution.

Theorem: Let the likelihood function of the observations $\mathbf{y}$ be determined by the normal distribution $\mathbf{y}|\boldsymbol{\beta},\tau \sim N(\mathbf{X}\boldsymbol{\beta},\tau^{-1}\mathbf{I})$ under the condition that $\boldsymbol{\beta}$ and τ are given. Let the prior distribution of the parameters $\boldsymbol{\beta}$ and τ have the normal-gamma distribution $\boldsymbol{\beta},\tau \sim NG(\boldsymbol{\mu},\mathbf{V},b,p)$, then the posterior distribution of $\boldsymbol{\beta}$ and τ is also normal-gamma

$$\boldsymbol{\beta},\tau|\mathbf{y} \sim NG(\boldsymbol{\mu}_o,\mathbf{V}_o,b_o,p_o)$$

with the parameters

$$\begin{aligned}
\boldsymbol{\mu}_o &= (\mathbf{X}'\mathbf{X}+\mathbf{V}^{-1})^{-1}(\mathbf{X}'\mathbf{y}+\mathbf{V}^{-1}\boldsymbol{\mu})\\
\mathbf{V}_o &= (\mathbf{X}'\mathbf{X}+\mathbf{V}^{-1})^{-1}\\
b_o &= (2b+(\boldsymbol{\mu}-\boldsymbol{\mu}_o)'\mathbf{V}^{-1}(\boldsymbol{\mu}-\boldsymbol{\mu}_o) + (\mathbf{y}-\mathbf{X}\boldsymbol{\mu}_o)'(\mathbf{y}-\mathbf{X}\boldsymbol{\mu}_o))/2\\
p_o &= (n+2p)/2. \qquad (224.10)
\end{aligned}$$

Proof: By multiplying the prior density of $\boldsymbol{\beta}$ and τ from (224.8) by the likelihood function (224.3) after substituting (224.6) we obtain with Bayes' theorem (211.1) the posterior density of $\boldsymbol{\beta}$ and τ as

$$\begin{aligned}
p(\boldsymbol{\beta},\tau|\mathbf{y}) &\propto \tau^{u/2+p-1}\exp\{-\frac{\tau}{2}[2b+(\boldsymbol{\beta}-\boldsymbol{\mu})'\mathbf{V}^{-1}(\boldsymbol{\beta}-\boldsymbol{\mu})]\}\\
&\quad \tau^{n/2}\exp[-\frac{\tau}{2}(\mathbf{y}-\mathbf{X}\boldsymbol{\beta})'(\mathbf{y}-\mathbf{X}\boldsymbol{\beta})]\\
&\propto \tau^{n/2+p+u/2-1}\exp\{-\frac{\tau}{2}[2b+(\boldsymbol{\beta}-\boldsymbol{\mu})'\mathbf{V}^{-1}(\boldsymbol{\beta}-\boldsymbol{\mu})+(\mathbf{y}-\mathbf{X}\boldsymbol{\beta})'(\mathbf{y}-\mathbf{X}\boldsymbol{\beta})]\}.
\end{aligned}$$

Completing the squares on $\boldsymbol{\beta}$ gives

$$\begin{aligned}
&2b + \boldsymbol{\beta}'(\mathbf{X}'\mathbf{X}+\mathbf{V}^{-1})\boldsymbol{\beta} - 2\boldsymbol{\beta}'(\mathbf{X}'\mathbf{y}+\mathbf{V}^{-1}\boldsymbol{\mu}) + \mathbf{y}'\mathbf{y} + \boldsymbol{\mu}'\mathbf{V}^{-1}\boldsymbol{\mu}\\
&= 2b + \mathbf{y}'\mathbf{y} + \boldsymbol{\mu}'\mathbf{V}^{-1}\boldsymbol{\mu} + (\boldsymbol{\beta}-\boldsymbol{\mu}_o)'(\mathbf{X}'\mathbf{X}+\mathbf{V}^{-1})(\boldsymbol{\beta}-\boldsymbol{\mu}_o) - \boldsymbol{\mu}_o'(\mathbf{X}'\mathbf{X}+\mathbf{V}^{-1})\boldsymbol{\mu}_o
\end{aligned}$$

with

$$\boldsymbol{\mu}_o = (\mathbf{X}'\mathbf{X}+\mathbf{V}^{-1})^{-1}(\mathbf{X}'\mathbf{y}+\mathbf{V}^{-1}\boldsymbol{\mu}).$$

The posterior density follows with

$$\begin{aligned}
p(\boldsymbol{\beta},\tau|\mathbf{y}) &\propto \tau^{u/2+(n+2p)/2-1}\exp\{-\frac{\tau}{2}[2b+\mathbf{y}'\mathbf{y}+\boldsymbol{\mu}'\mathbf{V}^{-1}\boldsymbol{\mu}-\boldsymbol{\mu}_o'(\mathbf{X}'\mathbf{X}+\mathbf{V}^{-1})\boldsymbol{\mu}_o\\
&\quad + (\boldsymbol{\beta}-\boldsymbol{\mu}_o)'(\mathbf{X}'\mathbf{X}+\mathbf{V}^{-1})(\boldsymbol{\beta}-\boldsymbol{\mu}_o)]\}.
\end{aligned}$$

Furthermore we have

$$2b + \mathbf{y}'\mathbf{y} + \boldsymbol{\mu}'\mathbf{V}^{-1}\boldsymbol{\mu} - \boldsymbol{\mu}_o'(\mathbf{X}'\mathbf{X}+\mathbf{V}^{-1})\boldsymbol{\mu}_o$$
$$= 2b + \mathbf{y}'\mathbf{y} + \boldsymbol{\mu}'\mathbf{V}^{-1}\boldsymbol{\mu} - 2\boldsymbol{\mu}_o'(\mathbf{X}'\mathbf{y}+\mathbf{V}^{-1}\boldsymbol{\mu}) + \boldsymbol{\mu}_o'(\mathbf{X}'\mathbf{X}+\mathbf{V}^{-1})\boldsymbol{\mu}_o$$
$$= 2b + (\boldsymbol{\mu}-\boldsymbol{\mu}_o)'\mathbf{V}^{-1}(\boldsymbol{\mu}-\boldsymbol{\mu}_o) + (\mathbf{y}-\mathbf{X}\boldsymbol{\mu}_o)'(\mathbf{y}-\mathbf{X}\boldsymbol{\mu}_o). \qquad (224.11)$$

After substituting this result we recognize $p(\boldsymbol{\beta},\tau|\mathbf{y})$ with (A23.1) as the density of the normal-gamma distribution with the parameters given in (224.10). □

If as a prior density for the parameters $\boldsymbol{\beta}$ and τ the normal-gamma distribution

$$\boldsymbol{\beta},\tau \sim NG(\boldsymbol{\mu},\mathbf{V},b,p) \qquad (224.12)$$

is chosen, the parameters $\boldsymbol{\mu}$, $\mathbf{V}$, b, p of the normal-gamma distribution need to be specified. This is readily done, if the expected value $E(\boldsymbol{\beta})$ of $\boldsymbol{\beta}$ and its covariance matrix $D(\boldsymbol{\beta})$ are given with

$$E(\boldsymbol{\beta}) = \boldsymbol{\mu}_p \quad \text{and} \quad D(\boldsymbol{\beta}) = \boldsymbol{\Sigma}_\beta \qquad (224.13)$$

and equivalently for σ^2 instead of τ with

$$E(\sigma^2) = \sigma_p^2 \quad \text{and} \quad V(\sigma^2) = V_{\sigma^2}. \qquad (224.14)$$

According to (A23.3) the marginal distribution of $\boldsymbol{\beta}$ in (224.12) is the multivariate t-distribution

$$\boldsymbol{\beta} \sim t(\boldsymbol{\mu},b\mathbf{V}/p,2p) \qquad (224.15)$$

with the expected value and the covariance matrix from (A22.7)

$$E(\boldsymbol{\beta}) = \boldsymbol{\mu} \quad \text{and} \quad D(\boldsymbol{\beta}) = b(p-1)^{-1}\mathbf{V}. \qquad (224.16)$$

The marginal distribution of τ in (224.12) is according to (A23.4)

$$\tau \sim G(b,p) \qquad (224.17)$$

and for $\sigma^2=1/\tau$ according to (A13.1) the inverted gamma distribution

$$\sigma^2 \sim IG(b,p) \qquad (224.18)$$

with the expected value and variance from (A13.2)

$$E(\sigma^2) = b/(p-1) \quad \text{and} \quad V(\sigma^2) = b^2/((p-1)^2(p-2)). \qquad (224.19)$$

Substituting (224.16) and (224.19) in (224.13) and (224.14) gives

$$\boldsymbol{\mu} = \boldsymbol{\mu}_p,\ \mathbf{V} = \boldsymbol{\Sigma}_\beta/\sigma_p^2,\ p = (\sigma_p^2)^2/V_{\sigma^2} + 2,\ b = (p-1)\sigma_p^2, \qquad (224.20)$$

which determines the parameters of the prior density (224.12) by means of the expected values and the variances and covariances of $\boldsymbol{\beta}$ and σ^2. The result for the parameter $\mathbf{V}$

could be expected, since $\boldsymbol{\beta}$ under the condition that τ is given has the covariance matrix $\boldsymbol{\Sigma}=\tau^{-1}\mathbf{V}=\sigma^2\mathbf{V}$, as can be seen from (A23.2).

We will apply the results of (224.10) and (224.20) to the special problem of extending the Example 1 of Section 211 such that both the expected value and the variance of the observations are unknown.

Example 1: We assume $\mathbf{X}=[1,\ldots,1]'$ and $\boldsymbol{\beta}=\mu$ in (224.10). This gives with

$$(\mathbf{y}-\mathbf{X}\boldsymbol{\beta})'(\mathbf{y}-\mathbf{X}\boldsymbol{\beta}) = \sum_{i=1}^{n}(y_i-\mu)^2$$

instead of (224.3) the likelihood function

$$p(\mathbf{y}|\mu,\sigma^2) = (2\pi\sigma^2)^{-n/2}\exp[-\frac{1}{2\sigma^2}\sum_{i=1}^{n}(y_i-\mu)^2], \tag{224.21}$$

which has the functional form of the likelihood function (211.10) of Example 1 of Section 211. However, both μ and σ^2 are unknown parameters in (224.21). As prior distribution for μ and $\tau=1/\sigma^2$ the normal-gamma distribution is assumed

$$\mu,\tau \sim NG(\bar{\mu},V,b,p), \tag{224.22}$$

whose parameters shall be determined by prior information on the expected values and variances of μ and σ^2. Hence,

$$E(\mu) = \mu_p,\quad V(\mu) = \sigma_\mu^2$$
$$E(\sigma^2) = \sigma_p^2,\quad V(\sigma^2) = V_{\sigma^2} \tag{224.23}$$

so that with (224.20) the parameters in (224.22) are determined by

$$\bar{\mu} = \mu_p,\quad V = \sigma_\mu^2/\sigma_p^2,\quad p = (\sigma_p^2)^2/V_{\sigma^2} + 2,\quad b = (p-1)\sigma_p^2. \tag{224.24}$$

The posterior distribution of μ and τ is given by

$$\mu,\tau|\mathbf{y} \sim NG(\mu_0,V_0,b_0,p_0) \tag{224.25}$$

with the parameters defined in (224.10). The marginal posterior distribution for μ follows from (A23.3) by the t-distribution

$$\mu|\mathbf{y} \sim t(\mu_0,b_0V_0/p_0,2p_0) \tag{224.26}$$

with the expected value and variance from (A22.7)

$$E(\mu) = \mu_0 \quad\text{and}\quad V(\mu) = b_0V_0/(p_0-1).$$

With (224.10) we find

$$E(\mu) = \mu_0 = (n+V^{-1})^{-1}(\sum_{i=1}^{n}y_i+V^{-1}\bar{\mu}) \tag{224.27}$$

and

$$V(\mu) = \frac{1}{n+2p-2}\,[2b+(\bar{\mu}-\mu_o)^2V^{-1}+\sum_{i=1}^{n}(y_i-\mu_o)^2](n+V^{-1})^{-1}. \qquad (224.28)$$

By substituting $\hat{\mu}$ from (211.11) we obtain with (224.24)

$$E(\mu) = \mu_o = \frac{n\hat{\mu}\sigma_\mu^2+\mu_p\sigma_p^2}{n\sigma_\mu^2+\sigma_p^2} = \frac{\hat{\mu}(\sigma_p^2/n)^{-1}+\mu_p(\sigma_\mu^2)^{-1}}{(\sigma_p^2/n)^{-1}+(\sigma_\mu^2)^{-1}} \qquad (224.29)$$

and

$$V(\mu) = [2+2(\sigma_p^2)^2/V_{\sigma^2}+(\mu_p-\mu_o)^2/\sigma_\mu^2+\sum_{i=1}^{n}(y_i-\mu_o)^2/\sigma_p^2]$$
$$[(n+2(\sigma_p^2)^2/V_{\sigma^2}+2)((\sigma_p^2/n)^{-1}+(\sigma_\mu^2)^{-1})]^{-1}. \qquad (224.30)$$

The variance of μ_p is σ_μ^2 and the variance of $\hat{\mu}$ resulting from the prior information is σ_p^2/n. The expected value $E(\mu)$ of μ is therefore obtained with (224.29) as the weighted mean of $\hat{\mu}$ and μ_p with $(\sigma_p^2/n)^{-1}$ and $(\sigma_\mu^2)^{-1}$ serving as weights. The weight $(V(\mu))^{-1}$ of μ from (224.30) results from the sum of the two weights modified by a factor. The results are therefore similar to (211.18) and (211.19).

Because of (224.25) the marginal posterior distribution of τ is obtained with (A23.4) by the gamma distribution

$$\tau|\mathbf{y} \sim G(b_o,p_o)$$

or with (A13.1) for σ^2 by the inverted gamma distribution

$$\sigma^2|\mathbf{y} \sim IG(b_o,p_o). \qquad (224.31)$$

Thus, we have with (A13.2)

$$E(\sigma^2) = b_o/(p_o-1) \quad \text{and} \quad V(\sigma^2) = b_o^2/((p_o-1)^2(p_o-2))$$

or with (224.10)

$$E(\sigma^2) = \frac{1}{n+2p-2}\,(2b+(\bar{\mu}-\mu_o)^2V^{-1}+\sum_{i=1}^{n}(y_i-\mu_o)^2) \qquad (224.32)$$

and

$$V(\sigma^2) = \frac{2(E(\sigma^2))^2}{n+2p-4}\,. \qquad (224.33)$$

By substituting (224.24) we finally obtain

$$E(\sigma^2) = [2\sigma_p^2+2(\sigma_p^2)^3/V_{\sigma^2} + (\mu_p-\mu_o)^2\sigma_p^2/\sigma_\mu^2+\sum_{i=1}^{n}(y_i-\mu_o)^2]$$
$$[n+2(\sigma_p^2)^2/V_{\sigma^2}+2]^{-1} \qquad (224.34)$$

and

$$V(\sigma^2) = \frac{2(E(\sigma^2))^2}{n+2(\sigma_p^2)^2/V_{\sigma^2}} . \tag{224.35}$$

Δ

225 Constraints for Parameters

In a latter section constraints will be imposed on the values which the parameters $\boldsymbol{\theta}$ can take. A logical way to proceed would be to choose the prior distribution for $\boldsymbol{\theta}$ such that the constraints are fulfilled. For instance, a linearly constrained least squares estimator for the parameters of a linear model may be derived with a limiting process by a sequence of Bayes estimators with a corresponding sequence of prior distributions (Pilz 1983, p.82). On the other hand, we may solve the unconstrained problem first and then impose the constraints on the posterior distribution. Both methods give identical results, as shown in the following

Theorem: Let C be a constraint such that

$$\boldsymbol{\theta} \in \Theta_c \quad \text{with} \quad \Theta_c \subset \Theta ,$$

i.e. Θ_c is a subspace of the parameter space Θ. Let $\boldsymbol{\theta}_r$ be a subset of $\boldsymbol{\theta}$ in the space Θ_r with $\Theta_r \subset \Theta_c$. Then the posterior distribution of $\boldsymbol{\theta}_r$ given the constraint C is equal to the posterior distribution for $\boldsymbol{\theta}_r$ without the constraint C multiplied by a factor depending on C

$$p(\boldsymbol{\theta}_r|C,\mathbf{y}) = p(\boldsymbol{\theta}_r|\mathbf{y}) \frac{P(C|\boldsymbol{\theta}_r,\mathbf{y})}{P(C|\mathbf{y})} . \tag{225.1}$$

Proof: From the definition of the conditional probability $P(A|B,D)$ of the event A, given that the events B and D have occurred, we obtain (Koch 1988a, p.93)

$$P(A|B,D) = \frac{P(A \cap B \cap D)}{P(B \cap D)} = \frac{P(A \cap B \cap D)/P(D)}{P(B \cap D)/P(D)} = \frac{P(A \cap B|D)}{P(B|D)} .$$

Furthermore,

$$P(B|A,D) = \frac{P(A \cap B \cap D)}{P(A \cap D)} = \frac{P(A \cap B \cap D)/P(D)}{P(A \cap D)/P(D)} = \frac{P(A \cap B|D)}{P(A|D)}$$

and by substituting this result in the first expression we find

$$P(A|B,D) = P(A|D)P(B|A,D)/P(B|D) . \tag{225.2}$$

Thus,

$$P(\boldsymbol{\theta}_r \in \Theta_r | C, \mathbf{y}) = P(\boldsymbol{\theta}_r \in \Theta_r | \mathbf{y}) P(C | \boldsymbol{\theta}_r \in \Theta_r, \mathbf{y}) / P(C | \mathbf{y}).$$

The probability for $\boldsymbol{\theta}_r \in \Theta_r$ can be thought of as resulting from a cumulative distribution function, so that the posterior distribution for $\boldsymbol{\theta}_r$ given the constraint C follows with

$$p(\boldsymbol{\theta}_r | C, \mathbf{y}) = p(\boldsymbol{\theta}_r | \mathbf{y}) \frac{P(C | \boldsymbol{\theta}_r, \mathbf{y})}{P(C | \mathbf{y})} \; . \qquad \square$$

23 Point Estimation

231 Quadratic Loss

Bayes' theorem (211.1) leads to the posterior distribution $p(\boldsymbol{\theta}|\mathbf{y})$ of the unknown parameters $\boldsymbol{\theta}$ given the data $\mathbf{y}$. All inferential problems concerning the paramters $\boldsymbol{\theta}$ can now be solved by means of $p(\boldsymbol{\theta}|\mathbf{y})$. Based on this distribution we will estimate the unknown parameters, establish confidence regions for them and test hypotheses concerning the parameters.

We start with the estimation, also called *point estimation*, in contrast to the estimation of confidence regions. The estimates $\hat{\boldsymbol{\theta}}$ of $\boldsymbol{\theta}$ shall be determined by the observations $\mathbf{y}$, hence $\hat{\boldsymbol{\theta}}=\hat{\boldsymbol{\theta}}(\mathbf{y})$. Based on the posterior distribution $p(\boldsymbol{\theta}|\mathbf{y})$ it has to be decided which are the estimates $\hat{\boldsymbol{\theta}}$ of the values of the parameters $\boldsymbol{\theta}$. The estimation may be therefore viewed as a statistical decision problem. With each decision ending in an estimate $\hat{\boldsymbol{\theta}}$ of the parameters $\boldsymbol{\theta}$ a *loss* $L(\boldsymbol{\theta},\hat{\boldsymbol{\theta}})$ is connected. The *posterior expected loss* of the estimation is defined by the expected value of the loss computed by the posterior density

$$E(L(\boldsymbol{\theta},\hat{\boldsymbol{\theta}})) = \int_{\Theta} L(\boldsymbol{\theta},\hat{\boldsymbol{\theta}})\ p(\boldsymbol{\theta}|\mathbf{y})d\boldsymbol{\theta}. \tag{231.1}$$

A *Bayes estimator* is now simply found by minimizing the posterior expected loss. In general we use the

Definition: A decision, which minimizes the posterior expected loss, is called a *Bayes rule.* (231.2)

In order to derive a Bayes estimator by means of the Bayes rule, we have to specify the loss. It is very simple to work with the quadratic loss defined by means of the error $\boldsymbol{\theta}-\hat{\boldsymbol{\theta}}$ of the estimate $\hat{\boldsymbol{\theta}}$

$$L(\boldsymbol{\theta},\hat{\boldsymbol{\theta}}) = (\boldsymbol{\theta}-\hat{\boldsymbol{\theta}})'\mathbf{P}(\boldsymbol{\theta}-\hat{\boldsymbol{\theta}}), \tag{231.3}$$

where $\mathbf{P}$ is assumed to be a given positive definite matrix of constants serving as a weight matrix. To compute the posterior expected loss we use the identity

$$\begin{aligned} E[(\boldsymbol{\theta}-\hat{\boldsymbol{\theta}})'\mathbf{P}(\boldsymbol{\theta}-\hat{\boldsymbol{\theta}})] &= E\{[\boldsymbol{\theta}-E(\boldsymbol{\theta})-(\hat{\boldsymbol{\theta}}-E(\boldsymbol{\theta}))]'\mathbf{P}[\boldsymbol{\theta}-E(\boldsymbol{\theta})-(\hat{\boldsymbol{\theta}}-E(\boldsymbol{\theta}))]\} \\ &= E[(\boldsymbol{\theta}-E(\boldsymbol{\theta}))'\mathbf{P}(\boldsymbol{\theta}-E(\boldsymbol{\theta}))] + (\hat{\boldsymbol{\theta}}-E(\boldsymbol{\theta}))'\mathbf{P}(\hat{\boldsymbol{\theta}}-E(\boldsymbol{\theta})) \end{aligned} \tag{231.4}$$

because of

$$E[(\boldsymbol{\theta}-E(\boldsymbol{\theta}))']\mathbf{P}(\hat{\boldsymbol{\theta}}-E(\boldsymbol{\theta})) = 0 \quad \text{with} \quad E[\boldsymbol{\theta}-E(\boldsymbol{\theta})] = \mathbf{0}.$$

The first term of (231.4) does not depend on $\hat{\boldsymbol{\theta}}$, while the second term is minimum for

$$\hat{\boldsymbol{\theta}}_B = E(\boldsymbol{\theta}), \tag{231.5}$$

since **P** is positive definite. Thus, $\hat{\boldsymbol{\theta}}_B$ is the *Bayes estimate*, when the loss function is quadratic. The estimate is defined by the expected value for $\boldsymbol{\theta}$ computed with the posterior density

$$\hat{\boldsymbol{\theta}}_B = \int_{\Theta} \boldsymbol{\theta} p(\boldsymbol{\theta}|\mathbf{y})d\boldsymbol{\theta}. \tag{231.6}$$

In the following we will mostly apply this Bayes estimate $\hat{\boldsymbol{\theta}}_B$. To express its accuracy, we introduce the covariance matrix $\boldsymbol{\Sigma}_{\boldsymbol{\theta}}$ of the estimate $\hat{\boldsymbol{\theta}}_B$ in analogy to (231.5) and (231.6) by

$$\boldsymbol{\Sigma}_{\boldsymbol{\theta}} = E[(\boldsymbol{\theta}-E(\boldsymbol{\theta}))(\boldsymbol{\theta}-E(\boldsymbol{\theta}))'] = \int_{\Theta} (\boldsymbol{\theta}-E(\boldsymbol{\theta}))(\boldsymbol{\theta}-E(\boldsymbol{\theta}))'p(\boldsymbol{\theta}|\mathbf{y})d\boldsymbol{\theta}. \tag{231.7}$$

By substituting (231.7) in (231.4) we find (Koch 1988a, p.156)

$$E[(\boldsymbol{\theta}-E(\boldsymbol{\theta}))'\mathbf{P}(\boldsymbol{\theta}-E(\boldsymbol{\theta}))] = E\{tr[\mathbf{P}(\boldsymbol{\theta}-E(\boldsymbol{\theta}))(\boldsymbol{\theta}-E(\boldsymbol{\theta}))']\} = tr\mathbf{P}\boldsymbol{\Sigma}_{\boldsymbol{\theta}},$$

so that the posterior expected loss for the Bayes estimate (231.5) is obtained by

$$E((\boldsymbol{\theta}-\hat{\boldsymbol{\theta}}_B)'\mathbf{P}(\boldsymbol{\theta}-\hat{\boldsymbol{\theta}}_B)) = tr\mathbf{P}\boldsymbol{\Sigma}_{\boldsymbol{\theta}}. \tag{231.8}$$

Example 1: The expected value (211.18) of the parameter μ of Example 1 of Section 211 gives the Bayes estimate $\hat{\mu}_B$ of μ, hence $\hat{\mu}_B=E(\mu)$ and the variance $V(\mu)$ from (211.19) leads to the variance σ^2_{μ} of $\hat{\mu}_B$, thus $\sigma^2_{\mu}=V(\mu)$. The same holds true for the expected value and the variance (222.16) of μ of Example 3 of Section 222 and for the expected values (224.29) and (224.34) and the variances (224.30) and (224.35) of the parameters μ and σ^2 of Example 1 of Section 224. **Δ**

232 Different Estimators

Different loss functions lead to different estimators. Let $\hat{\boldsymbol{\theta}}$ be the estimate of $\boldsymbol{\theta}$ with $\boldsymbol{\theta}=(\theta_i)$, $\hat{\boldsymbol{\theta}}=(\hat{\theta}_i)$ and $i\in\{1,\ldots,u\}$. If the absolute error $|\theta_i-\hat{\theta}_i|$ of the estimate $\hat{\theta}_i$ is chosen as a loss, we minimize the posterior expected loss

$$E(L(\theta_i,\hat{\theta}_i)) = \int_{\Theta} |\theta_i-\hat{\theta}_i|p(\boldsymbol{\theta}|\mathbf{y})d\boldsymbol{\theta}. \tag{232.1}$$

Let the domain of the integration, the parameter space Θ, be defined by the inequality

$$\theta_{oi} < \theta_i < \theta_{1i} \quad \text{for} \quad i\in\{1,\ldots,u\},$$

so that we obtain, since $|\theta_i - \hat{\theta}_i|$ is positive,

$$E(L(\theta_i,\hat{\theta}_i)) = \int_{\theta_{ou}}^{\hat{\theta}_u} \ldots \int_{\theta_{o1}}^{\hat{\theta}_1} (\hat{\theta}_i - \theta_i)\, p(\boldsymbol{\theta}|\mathbf{y})\, d\theta_1 \ldots d\theta_u$$

$$+ \int_{\hat{\theta}_u}^{\theta_{1u}} \ldots \int_{\hat{\theta}_1}^{\theta_{11}} (\theta_i - \hat{\theta}_i)\, p(\boldsymbol{\theta}|\mathbf{y})\, d\theta_1 \ldots d\theta_u$$

$$= \hat{\theta}_i P(\hat{\boldsymbol{\theta}}|\mathbf{y}) - \int_{\theta_{ou}}^{\hat{\theta}_u} \ldots \int_{\theta_{o1}}^{\hat{\theta}_1} \theta_i p(\boldsymbol{\theta}|\mathbf{y})\, d\theta_1 \ldots d\theta_u$$

$$+ \int_{\hat{\theta}_u}^{\theta_{1u}} \ldots \int_{\hat{\theta}_1}^{\theta_{11}} \theta_i p(\boldsymbol{\theta}|\mathbf{y})\, d\theta_1 \ldots d\theta_u - \hat{\theta}_i(1 - P(\hat{\boldsymbol{\theta}}|\mathbf{y})), \qquad (232.2)$$

where

$$P(\hat{\boldsymbol{\theta}}|\mathbf{y}) = \int_{\theta_{ou}}^{\hat{\theta}_u} \ldots \int_{\theta_{o1}}^{\hat{\theta}_1} p(\boldsymbol{\theta}|\mathbf{y})\, d\theta_1 \ldots d\theta_u$$

denotes the cumulative posterior distribution function. To find the minimum of (232.2), we differentiate with respect to $\hat{\theta}_i$ and set the derivative equal to zero. We obtain, since the results of the differentiation with respect to the limits $\hat{\theta}_i$ of the integrals in (232.2) cancel,

$$\partial E(L(\theta_i,\hat{\theta}_i))/\partial\hat{\theta}_i = P(\hat{\boldsymbol{\theta}}|\mathbf{y}) - 1 + P(\hat{\boldsymbol{\theta}}|\mathbf{y}) = 0$$

or

$$P(\hat{\boldsymbol{\theta}}|\mathbf{y}) = \frac{1}{2} \, . \qquad (232.3)$$

Thus, the Bayes estimate $\hat{\boldsymbol{\theta}}$ in case of the absolute errors $|\theta_i - \hat{\theta}_i|$ as a loss is the *median* of the posterior density function, that is the value $\hat{\boldsymbol{\theta}}$ for which according to (232.3) the cumulative distribution function is equal to 1/2. The median (232.3) minimizes (232.1), since the second derivative $\partial^2 E(L(\theta_i,\hat{\theta}_i))/\partial\hat{\theta}_i^2$ is positive.

In analogy to the standard statistical techniques the ***generalized maximum likelihood estimate*** $\bar{\boldsymbol{\theta}}$ of the parameter vector $\boldsymbol{\theta}$ is the *mode* of the posterior density $p(\boldsymbol{\theta}|\mathbf{y})$, i.e. the value for $\boldsymbol{\theta}$ which maximizes $p(\boldsymbol{\theta}|\mathbf{y})$, hence

$$\bar{\boldsymbol{\theta}} = \sup_{\boldsymbol{\theta}} p(\boldsymbol{\theta}|\mathbf{y}) . \tag{232.4}$$

This estimate is also called the ***maximum a posteriori*** or ***MAP estimate***.

24 Confidence Regions

241 H.P.D. Region

If the posterior density function $p(\boldsymbol{\theta}|\mathbf{y})$ of the parameter vector $\boldsymbol{\theta}$ has been determined by Bayes' theorem (211.1), we may compute the probability that the vector $\boldsymbol{\theta}$ lies in a subspace Θ_s of the parameter space Θ with $\Theta_s \subset \Theta$ by

$$P(\boldsymbol{\theta} \in \Theta_s|\mathbf{y}) = \int_{\Theta_s} p(\boldsymbol{\theta}|\mathbf{y})d\boldsymbol{\theta}. \tag{241.1}$$

Often we are interested in finding the subspace, where most, for instance 95 per cent, of the probability mass is concentrated. Obviously there are an infinite number of ways to specify such a region. To obtain a unique solution, provided the density has only one modal value, i.e. one maximum, the region should be defined such that the density of every point inside is at least as large as for any point outside of it. A region with such a property is called a region of highest posterior density (H.P.D. region) or Bayesian confidence region. In addition it has the property that for a given probability mass it occupies the smallest possible volume in the parameter space (Box and Tiao 1973, p.123). The first property will be applied in the

Definition: Let $p(\boldsymbol{\theta}|\mathbf{y})$ be a unimodal posterior density function of the parameter vector $\boldsymbol{\theta}$. A subspace B of the parameter space of $\boldsymbol{\theta}$ is called *H.P.D. region*, *Bayesian confidence region* or just *confidence region* of content $1-\alpha$, if

$$P(\boldsymbol{\theta} \in B|\mathbf{y}) = 1-\alpha$$

and

$$p(\boldsymbol{\theta}_1|\mathbf{y}) \geq p(\boldsymbol{\theta}_2|\mathbf{y}) \quad \text{for} \quad \boldsymbol{\theta}_1 \in B, \quad \boldsymbol{\theta}_2 \notin B. \tag{241.2}$$

As in the standard statistical techniques we will use small values of α, say α=0.01, α=0.05 or α=0.1. In general we set α=0.05, so that a 95 per cent confidence region is defined.

242 Boundary of a Confidence Region

After having established a confidence region for the parameters $\boldsymbol{\theta}$ the question may arise, whether a particular value $\boldsymbol{\theta}_0$ of the parameter vector lies inside or outside the confidence region. The event $\boldsymbol{\theta}\in B|\mathbf{y}$ of (241.2) is equivalent to $p(\boldsymbol{\theta}|\mathbf{y})>b$, where b is a constant and equal to the posterior density at the boundary of the confidence region

$$b = p(\boldsymbol{\theta}_b|\mathbf{y}) \tag{242.1}$$

with $\boldsymbol{\theta}_b$ denoting a point at the boundary. Hence, we define instead of (241.2)

$$P(\boldsymbol{\theta} \in B|\mathbf{y}) = P(p(\boldsymbol{\theta}|\mathbf{y}) > b) = 1-\alpha . \tag{242.2}$$

A particular value $\boldsymbol{\theta}_o$ of $\boldsymbol{\theta}$ lies inside the confidence region, if

$$P(p(\boldsymbol{\theta}|\mathbf{y}) > p(\boldsymbol{\theta}_o|\mathbf{y})) < 1-\alpha \tag{242.3}$$

or if the inequality is fulfilled

$$p(\boldsymbol{\theta}_o|\mathbf{y}) > b . \tag{242.4}$$

The last inequality represents a very simple way of checking whether $\boldsymbol{\theta}_o$ lies inside the confidence region.

In the expressions (242.2) and (242.3) the density function $p(\boldsymbol{\theta}|\mathbf{y})$ has the meaning of a random variable.

Example 1: Let us assume the posterior density $p(\boldsymbol{\theta}|\mathbf{y})$ of the u×1 parameter vector $\boldsymbol{\theta}$ is given by the multivariate t-distribution $t(\boldsymbol{\mu},\mathbf{N}^{-1},\nu)$, which is obtained by (A23.3) as a marginal distribution of the normal-gamma distribution, which in turn is defined according to (224.10). As can be seen from (A22.1), $p(\boldsymbol{\theta}|\mathbf{y})$ is a monotonically decreasing function of $(\boldsymbol{\theta}-\boldsymbol{\mu})'\mathbf{N}(\boldsymbol{\theta}-\boldsymbol{\mu})$. But this quadratic form divided by u has according to (A22.13) the F-distribution $F(u,\nu)$

$$(\boldsymbol{\theta}-\boldsymbol{\mu})'\mathbf{N}(\boldsymbol{\theta}-\boldsymbol{\mu})/u \sim F(u,\nu) .$$

Hence, according to (242.2), where the greater than sign has to be replaced by the smaller than sign, since $p(\boldsymbol{\theta}|\mathbf{y})$ decreases when $(\boldsymbol{\theta}-\boldsymbol{\mu})'\mathbf{N}(\boldsymbol{\theta}-\boldsymbol{\mu})$ increases, the confidence region of content $1-\alpha$ is defined by

$$P((\boldsymbol{\theta}-\boldsymbol{\mu})'\mathbf{N}(\boldsymbol{\theta}-\boldsymbol{\mu})/u < F_{1-\alpha;u,\nu}) = 1-\alpha . \tag{242.5}$$

$F_{1-\alpha;u,\nu}$ denotes the upper α-percentage point of the F-distribution with u and ν degrees of freedom (Koch 1988a, p.150). The boundary of the confidence region is obtained by

$$(\boldsymbol{\theta}-\boldsymbol{\mu})'\mathbf{N}(\boldsymbol{\theta}-\boldsymbol{\mu})/u = F_{1-\alpha;u,\nu} , \tag{242.6}$$

i.e. by a hyperellipsoid with the center at the point $\boldsymbol{\mu}$ (Koch 1988a, p.328).

To answer the question whether a particular point $\boldsymbol{\theta}_o$ lies inside the confidence region, we merely have to check according to (242.4), if

$$(\boldsymbol{\theta}_o-\boldsymbol{\mu})'\mathbf{N}(\boldsymbol{\theta}_o-\boldsymbol{\mu})/u < F_{1-\alpha;u,v} \tag{242.7}$$

is fulfilled. Δ

25 Hypothesis Testing

251 Different Hypotheses

Assumptions about the unknown parameters are called hypotheses. Let $\Theta_0 \subset \Theta$ and $\Theta_1 \subset \Theta$ be two subsets of the set Θ of the parameter vectors, the parameter space. Let Θ_0 and Θ_1 be disjoint, hence $\Theta_0 \cap \Theta_1 = \emptyset$. Then the assumption that the parameter vector $\boldsymbol{\theta}$ belongs to the subset Θ_0 is called the *null hypothesis* and the assumption that $\boldsymbol{\theta}$ belongs to Θ_1 the *alternative hypothesis*, hence

$$H_0 : \boldsymbol{\theta} \in \Theta_0 \quad \text{versus} \quad H_1 : \boldsymbol{\theta} \in \Theta_1 . \tag{251.1}$$

Frequently, Θ_1 is the complement of Θ_0

$$\Theta_1 = \Theta \backslash \Theta_0 . \tag{251.2}$$

The subset Θ_0 is assumed to contain more than one element, (251.1) is therefore called a *composite hypothesis* in contrast to the *simple hypothesis*

$$H_0 : \boldsymbol{\theta} = \boldsymbol{\theta}_0 \quad \text{versus} \quad H_1 : \boldsymbol{\theta} = \boldsymbol{\theta}_1 , \tag{251.3}$$

where the subsets Θ_0 and Θ_1 contain only the elements $\boldsymbol{\theta}_0$ and $\boldsymbol{\theta}_1$, respectively.

If the subset Θ_0 in (251.1) consists only of the point $\boldsymbol{\theta}_0$ and the subset Θ_1 is the complement of Θ_0 according to (251.2), we obtain the *point null hypothesis*

$$H_0 : \boldsymbol{\theta} = \boldsymbol{\theta}_0 \quad \text{versus} \quad H_1 : \boldsymbol{\theta} \neq \boldsymbol{\theta}_0 . \tag{251.4}$$

If not the parameter vector $\boldsymbol{\theta}$ itself but a linear combination $\mathbf{H}\boldsymbol{\theta}$ needs to be tested, where $\mathbf{H}$ denotes a given matrix of constants, we write instead of (251.1)

$$H_0 : \mathbf{H}\boldsymbol{\theta} \in \bar{\Theta}_0 \quad \text{versus} \quad H_1 : \mathbf{H}\boldsymbol{\theta} \in \bar{\Theta}_1 , \tag{251.5}$$

where $\bar{\Theta}_0$ and $\bar{\Theta}_1$ are subsets of the parameter space $\bar{\Theta}$ of the transformed parameters $\mathbf{H}\boldsymbol{\theta}$. The hypotheses (251.3) and (251.4) for the linearly transformed parameters are obtained accordingly. The latter hypothesis corresponds to the general linear hypothesis of a Gauss-Markoff model (Koch 1988a, p.307).

252 Hypothesis Testing by Confidence Regions

Let us assume that (251.2) in connection with (251.1) is valid, so that the hypothesis is given

$$H_0 : \boldsymbol{\theta} \in \Theta_0 \quad \text{versus} \quad H_1 : \boldsymbol{\theta} \in \Theta \backslash \Theta_0 . \tag{252.1}$$

To decide whether to accept or to reject the null hypothesis, we compute by means of the posterior density $p(\boldsymbol{\theta}|\mathbf{y})$ according to (241.1) the probability of $\boldsymbol{\theta} \in \Theta_0$ with

$$P(\boldsymbol{\theta} \in \Theta_0|\mathbf{y}) = \int_{\Theta_0} p(\boldsymbol{\theta}|\mathbf{y})d\boldsymbol{\theta}. \tag{252.2}$$

By following the argument of the standard statistical techniques we reject the null hypothesis of (252.1), if

$$P(\boldsymbol{\theta} \in \Theta_0|\mathbf{y}) > 1-\alpha, \tag{252.3}$$

where $1-\alpha$ denotes the content of the confidence region (241.2). This can be explained by assuming $\Theta \backslash \Theta_0$ as the region of rejection of the test. The probability of the Type I error of the standard statistical techniques, that is of the rejection of a true null hypothesis, is then defined to be α

$$P(\boldsymbol{\theta} \in \Theta \backslash \Theta_0|\mathbf{y}) = \alpha \quad \text{or} \quad P(\boldsymbol{\theta} \in \Theta_0|\mathbf{y}) = 1-\alpha.$$

If Θ_0 is now given such that (252.3) is valid, Θ_0 extends into the region of rejection and the null hypothesis has to be rejected.

We can use the relation (252.3) also for the test of the point null hypothesis (251.4). The shape of the subspace Θ_0 in (252.3) may be arbitrarily defined, so that we give it the shape of the confidence region B defined by (241.2). The test of the point null hypothesis (251.4) is then reduced to the check, whether the point $\boldsymbol{\theta}_0$ of (251.4) lies inside or outside the confidence region B. Thus, in case of testing

$$H_0 : \boldsymbol{\theta} = \boldsymbol{\theta}_0 \quad \text{versus} \quad H_1 : \boldsymbol{\theta} \neq \boldsymbol{\theta}_0$$

we reject the null hypothesis, if according to (242.4)

$$p(\boldsymbol{\theta}_0|\mathbf{y}) < b \tag{252.4}$$

is fulfilled. The constant b is defined with (242.1) and denotes the posterior density at the boundary of the confidence region B. If not the parameter vector $\boldsymbol{\theta}$ itself but a linear combination $\mathbf{H}\boldsymbol{\theta}$ of the parameters has to be tested, the posterior density for the linear combination is determined and with it the test runs off correspondingly.

The test procedure (252.4) can be substantiated by the fact that the posterior density contains the information on possible values of the parameters. If the value $\boldsymbol{\theta}_0$ lies in a region where the posterior density is low, we should not trust this value and should reject the hypothesis.

When applying this method of hypothesis testing, the prior distribution for the parameters should be smooth in the vicinity of the point $\boldsymbol{\theta}_0$, i.e. the prior density should not change much when using instead of $\boldsymbol{\theta}_o$ a point in the neighborhood of $\boldsymbol{\theta}_o$ (Lindley 1965, Part 2, p.61). This will be true for many applications and it will be always true in the case of noninformative priors. If the prior density changes rapidly in the vicinity of $\boldsymbol{\theta}_o$, or if special priors have to be associated with the hypotheses, the procedure of Section 254 should be applied. As will be seen when testing the point null hypothesis, this method has its deficiency, too, so that in the case of smooth priors it is recommended to test the point null hypotheses by means of confidence regions.

Hypothesis testing by means of confidence regions is equivalent to the test of a general hypothesis in the Gauss-Markoff model by the standard statistical techniques (Koch 1988a, p.331). Both methods therefore give identical results, if the posterior density for the unknown parameters or for functions of the parameters agrees with the density of the test statistic of the standard techniques.

Example 1: The confidence region of a parameter vector $\boldsymbol{\theta}$ with a multivariate t-distribution was determined in the Example 1 of Section 242. It was also checked by the inequality (242.7), whether a point $\boldsymbol{\theta}_o$ lies inside the confidence region. Hence, if we test

$$H_o : \boldsymbol{\theta} = \boldsymbol{\theta}_o \quad \text{versus} \quad H_1 : \boldsymbol{\theta} \neq \boldsymbol{\theta}_o ,$$

the null hypothesis H_o is accepted according to (252.4), if the inequality (242.7) is fulfilled. Δ

253 Posterior Probabilities of Hypotheses

Corresponding to the point estimation, the test of the hypothesis (251.1) can be viewed as a decision problem, to accept H_o or to accept H_1, which are two actions. Two states of nature are connected with the hypothesis, either H_o is true or H_1 is true, thus, we have a two-state-two-action problem. Correspondingly, the loss connected with an action H_i in a specific state $\boldsymbol{\theta} \in \Theta_i$ has to be defined for four values

$$L(\boldsymbol{\theta} \in \Theta_i, H_i) = 0 \quad \text{for} \quad i \in \{0,1\}$$

$$L(\boldsymbol{\theta} \in \Theta_i, H_j) \neq 0 \quad \text{for} \quad i,j \in \{0,1\},\ i \neq j . \tag{253.1}$$

Zero loss is assumed for the correct decision, accept H_o and H_1 if $\boldsymbol{\theta} \in \Theta_o$ and $\boldsymbol{\theta} \in \Theta_1$, respectively, and loss not equal to zero occurs for a wrong decision.

To reach a decision based on the Bayes rule (231.2), the posterior expected loss of the actions has to be computed. Hence, we need the posterior probability $P(H_o|\mathbf{y})$ for the hypothesis H_o and $P(H_1|\mathbf{y})$ for H_1. These probabilities are computed according to (241.1) by

$$P(H_i|\mathbf{y}) = P(\boldsymbol{\theta} \in \Theta_i|\mathbf{y}) = \int_{\Theta_i} p(\boldsymbol{\theta}|\mathbf{y})d\boldsymbol{\theta} \quad \text{for} \quad i \in \{0,1\}. \tag{253.2}$$

If (251.2) holds true, then

$$P(H_1|\mathbf{y}) = 1-P(H_o|\mathbf{y}). \tag{253.3}$$

The posterior expected loss for accepting H_o follows with

$$E(L|H_o) = P(H_o|\mathbf{y})L(\boldsymbol{\theta} \in \Theta_o,H_o) + P(H_1|\mathbf{y})L(\boldsymbol{\theta} \in \Theta_1,H_o)$$
$$= P(H_1|\mathbf{y})L(\boldsymbol{\theta} \in \Theta_1,H_o),$$

since $L(\boldsymbol{\theta} \in \Theta_o,H_o)=0$ from (253.1). Correspondingly

$$E(L|H_1) = P(H_o|\mathbf{y})L(\boldsymbol{\theta} \in \Theta_o,H_1).$$

By minimizing the expected loss we obtain the Bayes rule for accepting H_o or H_1, which says,

$$\text{if} \quad E(L|H_o) < E(L|H_1), \quad \text{accept} \quad H_o \tag{253.4}$$

otherwise accept H_1. Thus, for testing hypothesis (251.1)

$$H_o : \boldsymbol{\theta} \in \Theta_o \quad \text{versus} \quad H_1: \boldsymbol{\theta} \in \Theta_1$$

we find the Bayes rule,

$$\text{if} \quad \frac{P(H_o|\mathbf{y})L(\boldsymbol{\theta} \in \Theta_o,H_1)}{P(H_1|\mathbf{y})L(\boldsymbol{\theta} \in \Theta_1,H_o)} > 1, \quad \text{accept} \quad H_o. \tag{253.5}$$

In the following we will assign equal losses for the wrong decisions

$$L(\boldsymbol{\theta} \in \Theta_o,H_1) = L(\boldsymbol{\theta} \in \Theta_1,H_o)$$

and obtain instead of (253.5) the Bayes rule,

$$\text{if} \quad \frac{P(H_o|\mathbf{y})}{P(H_1|\mathbf{y})} > 1, \quad \text{accept} \quad H_o \tag{253.6}$$

otherwise accept H_1. The ratio $P(H_o|\mathbf{y})/P(H_1|\mathbf{y})$ in (253.6) is called the *posterior odds* for H_o to H_1, it is computed from (253.2).

If the subspace Θ_o in the hypothesis (251.1) shrinks to the point $\boldsymbol{\theta}_o$, so that in case (251.2) is valid, the point null hypothesis (251.4) is obtained, then $P(H_o|\mathbf{y})$ in (253.6) goes to zero, since $p(\boldsymbol{\theta}|\mathbf{y})$ in (253.2) is continuous. This, of course, is not correct, so that a different procedure, which will be presented in the next section, has to be applied for testing the point null hypothesis. However, if both subspaces Θ_o and Θ_1 in (251.1) shrink to the points $\boldsymbol{\theta}_o$ and $\boldsymbol{\theta}_1$, so that instead of the composite hypothesis (251.1) the simple hypothesis (251.3) is tested, the posterior odds in (253.6) are computed with

(253.2) by

$$\frac{P(H_o|\mathbf{y})}{P(H_1|\mathbf{y})} = \frac{\lim_{\Delta\Theta_o\to 0} \int_{\Delta\Theta_o} p(\boldsymbol{\theta}|\mathbf{y})d\boldsymbol{\theta}/\Delta\Theta_o}{\lim_{\Delta\Theta_1\to 0} \int_{\Delta\Theta_1} p(\boldsymbol{\theta}|\mathbf{y})d\boldsymbol{\theta}/\Delta\Theta_1} = \frac{p(\boldsymbol{\theta}_o|\mathbf{y})}{p(\boldsymbol{\theta}_1|\mathbf{y})}, \qquad (253.7)$$

where the domains $\Delta\Theta_o$ and $\Delta\Theta_1$ of both integrals consist of small spaces around the points $\boldsymbol{\theta}_o$ and $\boldsymbol{\theta}_1$. For testing the hypothesis (251.3)

$$H_o : \boldsymbol{\theta} = \boldsymbol{\theta}_o \quad \text{versus} \quad H_1 : \boldsymbol{\theta} = \boldsymbol{\theta}_1$$

we therefore apply the Bayes rule,

$$\text{if} \quad \frac{p(\boldsymbol{\theta}_o|\mathbf{y})}{p(\boldsymbol{\theta}_1|\mathbf{y})} > 1, \quad \text{accept} \quad H_o \qquad (253.8)$$

otherwise accept H_1.

With this result we are able to interpret the hypothesis testing by confidence regions. The simple hypothesis

$$H_o : \boldsymbol{\theta} = \boldsymbol{\theta}_o \quad \text{versus} \quad H_1 : \boldsymbol{\theta} = \boldsymbol{\theta}_b, \qquad (253.9)$$

where $\boldsymbol{\theta}_b$ denotes a point at the boundary of the confidence region, is tested by means of the Bayes rule (253.8),

$$\text{if} \quad \frac{p(\boldsymbol{\theta}_o|\mathbf{y})}{p(\boldsymbol{\theta}_b|\mathbf{y})} > 1, \quad \text{accept} \quad H_o. \qquad (253.10)$$

Because of $b=p(\boldsymbol{\theta}_b|\mathbf{y})$ from (242.1), (253.10) is identical with (252.4).

It is worth mentioning that for one-sided tests

$$H_o : \boldsymbol{\theta} \le \mathbf{0} \quad \text{versus} \quad H_1 : \boldsymbol{\theta} > \mathbf{0}$$

the Bayes rule (253.6) gives results which are in agreement with the results of the standard statistical techniques (Casella and Berger 1987).

Example 1: Let the observations y_i of Example 1 of Section 211 represent independent measurements of the length of a straight line. The prior information μ_p and σ^2_μ on the unknown expected value μ of y_i and its variance is given by μ_p=5319.0 cm and σ^2_μ=49.0 cm². For n=5 observations with variance σ^2=9.0 cm² we obtained from (211.11) $\hat{\mu}$=5332.0 cm and therefore from (211.18) and (211.19) the expected value and variance of μ

$$E(\mu) = 5331.54 \text{ cm} \quad \text{and} \quad V(\mu) = 1.74 \text{ cm}^2.$$

Hence, μ is normally distributed according to

$$\mu|\mathbf{y} \sim N(5331.54,\ 1.74).$$

We want to test the hypothesis

$$H_0 : \mu \leq 5330 \quad \text{versus} \quad H_1 : \mu > 5330.$$

It is a composite hypothesis of type (251.1), so that we apply the Bayes rule (253.6). If $F(x;0,1)$ denotes the cumulative distribution function of the standard normal distribution (Koch 1988a, p.127), we obtain with (253.2) the posterior probability of the null hypothesis given the data

$$P(H_0|\mathbf{y}) = F((5330-5331.54)/\sqrt{1.74};0,1) = 0.12$$

and with (253.3) the posterior probability of the alternative hypothesis

$$P(H_1|\mathbf{y}) = 0.88.$$

Thus, the posterior odds for H_0 result with

$$P(H_0|\mathbf{y})/P(H_1|\mathbf{y}) = 0.14,$$

so that the null hypothesis has to be rejected. This result, of course, could have been foreseen, since the normal distribution is symmetrical with the center at 5331.54 in our example. Δ

254 Special Priors for Hypotheses

The methods of hypothesis testing presented in the preceding Sections 252 and 253 are based on the posterior density $p(\boldsymbol{\theta}|\mathbf{y})$ of the parameters. There are applications, however, when specific prior probabilities can be associated with the hypotheses. To deal with such a situation, the prior density function $p(\boldsymbol{\theta})$ of the parameter vector $\boldsymbol{\theta}$ is conveniently defined by

$$p(\boldsymbol{\theta}) = p_0 h_0(\boldsymbol{\theta}) + (1-p_0)h_1(\boldsymbol{\theta}) \tag{254.1}$$

with p_0 being a constant and $h_0(\boldsymbol{\theta})$ and $h_1(\boldsymbol{\theta})$ density functions so that (211.6) is fulfilled. The density $h_0(\boldsymbol{\theta})$ is defined on the subspace Θ_0 of the null hypothesis of (251.1) and $h_1(\boldsymbol{\theta})$ on Θ_1. The distributions $h_0(\boldsymbol{\theta})$ and $h_1(\boldsymbol{\theta})$ describe how the prior probability mass is spread out over the space of the null hypothesis and the alternative hypothesis. Using (254.1) together with the likelihood function $p(\mathbf{y}|\boldsymbol{\theta})$ we compute with Bayes' theorem (211.4) the posterior density $p(\boldsymbol{\theta}|\mathbf{y})$ by

$$p(\boldsymbol{\theta}|\mathbf{y}) = c[p_0 h_0(\boldsymbol{\theta})+(1-p_0)h_1(\boldsymbol{\theta})]p(\mathbf{y}|\boldsymbol{\theta}) \tag{254.2}$$

with

$$c = 1/\int_{\Theta} [p_o h_o(\boldsymbol{\theta})+(1-p_o)h_1(\boldsymbol{\theta})]p(\mathbf{y}|\boldsymbol{\theta})d\boldsymbol{\theta}.$$

Substituting this result in (253.2) leads to the posterior probability $P(H_o|\mathbf{y})$ of the hypothesis H_o

$$P(H_o|\mathbf{y}) = c\int_{\Theta_o} p_o h_o(\boldsymbol{\theta})p(\mathbf{y}|\boldsymbol{\theta})d\boldsymbol{\theta} \tag{254.3}$$

and correspondingly to

$$P(H_1|\mathbf{y}) = c\int_{\Theta_1} (1-p_o)h_1(\boldsymbol{\theta})p(\mathbf{y}|\boldsymbol{\theta})d\boldsymbol{\theta}. \tag{254.4}$$

The decision of accepting or rejecting the hypothesis (251.1) is then based on (253.6).

We let the space Θ_o in (251.1) shrink to the point $\boldsymbol{\theta}_o$ by introducing the space $\Delta\Theta$ with a small volume around the point $\boldsymbol{\theta}_o$ and obtain

$$\lim_{\Delta\Theta\to 0} \int_{\Delta\Theta} p_o h_o(\boldsymbol{\theta})p(\mathbf{y}|\boldsymbol{\theta})d\boldsymbol{\theta} = p_o p(\mathbf{y}|\boldsymbol{\theta}_o), \tag{254.5}$$

since $p(\mathbf{y}|\boldsymbol{\theta})$ can be considered being constant in $\Delta\Theta$ and $h_o(\boldsymbol{\theta})$ is a density defined on $\Delta\Theta$. In addition we let Θ_1 shrink to $\boldsymbol{\theta}_1$, so that the hypothesis (251.3) is obtained. The posterior odds then follow with substituting (254.5) in (254.3) and with an equivalent substitution in (254.4) by

$$\frac{P(H_o|\mathbf{y})}{P(H_1|\mathbf{y})} = \frac{p_o p(\mathbf{y}|\boldsymbol{\theta}_o)}{(1-p_o)p(\mathbf{y}|\boldsymbol{\theta}_1)} .$$

For testing the hypothesis (251.3)

$$H_o : \boldsymbol{\theta} = \boldsymbol{\theta}_o \quad \text{versus} \quad H_1 : \boldsymbol{\theta} = \boldsymbol{\theta}_1$$

we therefore apply according to (253.6) the Bayes rule,

$$\text{if} \quad \frac{p_o p(\mathbf{y}|\boldsymbol{\theta}_o)}{(1-p_o)p(\mathbf{y}|\boldsymbol{\theta}_1)} > 1, \quad \text{accept } H_o. \tag{254.6}$$

Finally we let Θ_o in (251.1) shrink to $\boldsymbol{\theta}_o$ with (251.2) holding true, so that the point null hypothesis (251.4) is obtained. The posterior probability $P(H_o|\mathbf{y})$ of H_o then follows with (254.2) and (254.5) from (254.3) by

$$P(H_o|\mathbf{y}) = p_o p(\mathbf{y}|\boldsymbol{\theta}_o)/[p_o p(\mathbf{y}|\boldsymbol{\theta}_o)+(1-p_o)\int_{\Theta} h_1(\boldsymbol{\theta})p(\mathbf{y}|\boldsymbol{\theta})d\boldsymbol{\theta}]. \tag{254.7}$$

For testing the point null hypothesis (251.4)

$$H_o : \boldsymbol{\theta} = \boldsymbol{\theta}_o \quad \text{versus} \quad H_1 : \boldsymbol{\theta} \neq \boldsymbol{\theta}_o$$

we therefore obtain with (253.3) from (253.6) the Bayes rule,

$$\text{if} \quad \frac{p_o p(\mathbf{y}|\boldsymbol{\theta}_o)}{(1-p_o) \int_{\Theta} h_1(\boldsymbol{\theta}) p(\mathbf{y}|\boldsymbol{\theta}) d\boldsymbol{\theta}} > 1 \quad \text{accept} \quad H_o \tag{254.8}$$

otherwise accept H_1.

It should be mentioned that using (254.8) for testing a point null hypothesis gives results that may differ from those of the standard statistical tests. This effect appears, if the prior density $h_1(\boldsymbol{\theta})$ is spread out considerably because of a large variance for the prior information on $\boldsymbol{\theta}$. Then the likelihood function averaged by the integral in (254.8) over the space of the alternative hypothesis becomes smaller than the likelihood function $p(\mathbf{y}|\boldsymbol{\theta}_o)$ for the null hypothesis. Thus, H_o is accepted although the standard statistical test may reject it. This discrepancy was first pointed out by Lindley (1957), see also Berger (1985, p.151), and it is called Lindley's paradox (Berger and Sellke 1987; Zellner 1971, p.304). A criticism of the Bayes rule (254.8) can be found, for instance, in (Casella and Berger 1987; Shafer 1982).

Example 1: We want to test by means of (254.8) the point null hypothesis

$$H_o : \mu = \mu_o \quad \text{versus} \quad H_1 : \mu \neq \mu_o \tag{254.9}$$

for the parameter μ of the Example 1 of Section 211. The likelihood function $p(\mathbf{y}|\mu)$ is therefore given by (211.14). The density $h_1(\mu)$, which according to (254.1) spreads out the prior probability over the space of the alternative hypothesis, shall be given by (211.8). We therefore obtain the posterior odds $P(H_o|\mathbf{y})/P(H_1|\mathbf{y})$ for H_o to H_1 from (254.8) by

$$P(H_o|\mathbf{y})/P(H_1|\mathbf{y}) = \sqrt{2\pi}\, \sigma_\mu\, p_o \exp[-\frac{n}{2\sigma^2}(\mu_o-\hat{\mu})^2] /$$
$$\{(1-p_o) \int_{-\infty}^{\infty} \exp[-\frac{1}{2}(\frac{(\mu-\mu_p)^2}{\sigma_\mu^2} + \frac{n}{\sigma^2}(\mu-\hat{\mu})^2)]d\mu\}.$$

Because of (211.16) the integral with respect to μ can be represented by an integral over the density function (211.17) of a normal distribution of expected value (211.18) and variance (211.19). We thus obtain with (A11.2)

$$\int_{-\infty}^{\infty} \exp[-\frac{1}{2}(\frac{(\mu-\mu_p)^2}{\sigma_\mu^2} + \frac{n}{\sigma^2}(\mu-\hat{\mu})^2)]d\mu$$

$$= \sqrt{2\pi}\, (\frac{\sigma_\mu^2\sigma^2/n}{\sigma_\mu^2+\sigma^2/n})^{1/2} \exp\{-\frac{1}{2}[\frac{\sigma_\mu^2+\sigma^2/n}{\sigma_\mu^2\sigma^2/n}(\frac{\hat{\mu}^2\sigma_\mu^2+\mu_p^2\sigma^2/n}{\sigma_\mu^2+\sigma^2/n} - (\frac{\hat{\mu}\sigma_\mu^2+\mu_p\sigma^2/n}{\sigma_\mu^2+\sigma^2/n})^2)]\}$$

$$= \sqrt{2\pi}\,\left(\frac{\sigma_\mu^2\sigma^2/n}{\sigma_\mu^2+\sigma^2/n}\right)^{1/2} \exp\left[-\frac{(\hat{\mu}-\mu_p)^2}{2(\sigma_\mu^2+\sigma^2/n)}\right]$$

and finally

$$\frac{P(H_o|\mathbf{y})}{P(H_1|\mathbf{y})} = \frac{p_o}{1-p_o}\left(\frac{\sigma_\mu^2+\sigma^2/n}{\sigma^2/n}\right)^{1/2} \exp\left[-\frac{1}{2}\left(\frac{(\mu_o-\hat{\mu})^2}{\sigma^2/n} - \frac{(\hat{\mu}-\mu_p)^2}{\sigma_\mu^2+\sigma^2/n}\right)\right]. \qquad (254.10)$$

With large values of σ_μ^2 we can make the posterior odds arbitrarily large. As already mentioned, this is due to the fact that if we spread out the density $h_1(\mu)$ by means of σ_μ^2, the likelihood averaged by $h_1(\mu)$ in the integral of (254.8) becomes smaller than the likelihood function at the point of the null hypothesis. For a special case we substitute in (254.10)

$$\sigma_\mu^2 = \sigma^2,\ \mu_p=\mu_o,\ z = |\hat{\mu}-\mu_o|/(\sigma/\sqrt{n})$$

and find

$$\frac{P(H_o|\mathbf{y})}{P(H_1|\mathbf{y})} = \frac{p_o}{1-p_o}(n+1)^{1/2} \exp\left[-\frac{1}{2} z^2\left(\frac{n}{n+1}\right)\right]. \qquad (254.11)$$

The quantity z is the test statistic of the hypothesis (254.9) by the standard statistical techniques (cf. Bosch 1985, p.84). If we fix z at the value for rejecting H_o on a significance level of α, for instance $z=1.96$ for $\alpha=0.05$, we will still find values for n which are large enough to make $P(H_o|\mathbf{y})/P(H_1|\mathbf{y})>1$, so that H_o has to be accepted. This is Lindley's paradox. Δ

26 Predictive Analysis

261 Joint Conditional Density Function

In the following section and in latter sections we need joint conditional density functions of random vectors expressed by marginal conditional density functions. They are obtained by the definition (211.2) of a conditional density function. Let $\mathbf{x}_1$, $\mathbf{x}_2$ and $\mathbf{x}_3$ be vectors of random variables. Then the conditional density function of $\mathbf{x}_1$ given $\mathbf{x}_2$ and $\mathbf{x}_3$ follows from (211.2) with

$$p(\mathbf{x}_1|\mathbf{x}_2,\mathbf{x}_3) = \frac{p(\mathbf{x}_1,\mathbf{x}_2,\mathbf{x}_3)}{p(\mathbf{x}_2,\mathbf{x}_3)} = \frac{p(\mathbf{x}_1,\mathbf{x}_2,\mathbf{x}_3)/p(\mathbf{x}_3)}{p(\mathbf{x}_2,\mathbf{x}_3)/p(\mathbf{x}_3)} = \frac{p(\mathbf{x}_1,\mathbf{x}_2|\mathbf{x}_3)}{p(\mathbf{x}_2|\mathbf{x}_3)}$$

or

$$p(\mathbf{x}_1,\mathbf{x}_2|\mathbf{x}_3) = p(\mathbf{x}_1|\mathbf{x}_2,\mathbf{x}_3)\ p(\mathbf{x}_2|\mathbf{x}_3). \tag{261.1}$$

Thus, the joint conditional density function $p(\mathbf{x}_1,\mathbf{x}_2|\mathbf{x}_3)$ of $\mathbf{x}_1$ and $\mathbf{x}_2$ given $\mathbf{x}_3$ is obtained by the conditional density function of $\mathbf{x}_1$ given $\mathbf{x}_2$ and $\mathbf{x}_3$ and by the marginal conditional density function of $\mathbf{x}_2$ given $\mathbf{x}_3$.

262 Predictive Distribution

Collecting data or making measurements generally takes time and effort. It is therefore appropriate to look for ways of predicting observations. This may be interpreted as either interpolating given data or forecasting observations from given data. In either case unobserved data are predicted.

We start from the given observations $\mathbf{y}$, which were introduced in Section 211 as a function of the parameter vector $\boldsymbol{\theta}$. Let the posterior density function $p(\boldsymbol{\theta}|\mathbf{y})$ of the parameters $\boldsymbol{\theta}$ given the data y be known and let the vector $\mathbf{y}_u$ denote the vector of unobserved data. The joint probability density function $p(\mathbf{y}_u,\boldsymbol{\theta}|\mathbf{y})$ of $\mathbf{y}_u$ and $\boldsymbol{\theta}$ given the data $\mathbf{y}$ is then obtained with (261.1) by

$$p(\mathbf{y}_u,\boldsymbol{\theta}|\mathbf{y}) = p(\mathbf{y}_u|\boldsymbol{\theta},\mathbf{y})\ p(\boldsymbol{\theta}|\mathbf{y}), \tag{262.1}$$

where $p(\mathbf{y}_u|\boldsymbol{\theta},\mathbf{y})$ is the conditional probability density function of $\mathbf{y}_u$ given $\boldsymbol{\theta}$ and y. If the same distribution for $\mathbf{y}_u$ is assumed as for the data y, then $p(\mathbf{y}_u|\boldsymbol{\theta},\mathbf{y})$ is known. By computing from (262.1) the marginal distribution for $\mathbf{y}_u$ (Koch 1988a, p.105) we obtain

$$p(\mathbf{y}_u|\mathbf{y}) = \int_{\Theta} p(\mathbf{y}_u|\boldsymbol{\theta},\mathbf{y})p(\boldsymbol{\theta}|\mathbf{y})d\boldsymbol{\theta}, \tag{262.2}$$

where Θ again denotes the parameter space. The density (262.2) is the *predictive density function* of the unobserved data vector $\mathbf{y}_u$. Any predictive inference for the unobserved data $\mathbf{y}_u$ is solved by the distribution $p(\mathbf{y}_u|\mathbf{y})$.

Example 1: For the Example 1 of Section 211 we have assumed n independent observations $\mathbf{y}=[y_1,\ldots,y_n]'$, each being normally distributed with unknown expected value μ and known variance σ^2. We want to derive the predictive distribution for the unobserved data point y_u. By assuming the same distribution as for the observation y_i the conditional density $p(y_u|\mu,\mathbf{y})$ of y_u is obtained from (211.9) by

$$p(y_u|\mu,\mathbf{y}) \propto \exp[-\frac{1}{2\sigma^2}(y_u-\mu)^2]. \tag{262.3}$$

The prior density function $p(\mu)$ of the unknown parameter μ follows from the normal distribution (211.8) and the posterior distribution $p(\mu|\mathbf{y})$ from (211.17) to (211.19) by

$$p(\mu|\mathbf{y}) \propto \exp[-\frac{1}{2V(\mu)}(\mu-E(\mu))^2]. \tag{262.4}$$

The predictive density function $p(y_u|\mathbf{y})$ of y_u is therefore given with (262.2) by

$$p(y_u|\mathbf{y}) = \int_{-\infty}^{\infty} p(y_u|\mu,\mathbf{y})p(\mu|\mathbf{y})d\mu$$
$$\propto \int_{-\infty}^{\infty} \exp\{-\frac{1}{2}[\frac{1}{\sigma^2}(y_u-\mu)^2 + \frac{1}{V(\mu)}(\mu-E(\mu))^2]\}d\mu. \tag{262.5}$$

We complete the square on μ in the exponent of (262.5) and obtain

$$\frac{1}{\sigma^2}(y_u-\mu)^2 + \frac{1}{V(\mu)}(\mu-E(\mu))^2$$
$$= \frac{V(\mu)+\sigma^2}{\sigma^2V(\mu)}[\mu^2-2\mu\frac{y_uV(\mu)+E(\mu)\sigma^2}{V(\mu)+\sigma^2}] + \frac{y_u^2V(\mu)+(E(\mu))^2\sigma^2}{\sigma^2V(\mu)}$$
$$= \frac{V(\mu)+\sigma^2}{\sigma^2V(\mu)}[\mu-\frac{y_uV(\mu)+E(\mu)\sigma^2}{V(\mu)+\sigma^2}]^2 + \frac{(y_u-E(\mu))^2}{V(\mu)+\sigma^2}.$$

By substituting this result in (262.5) we recognize that because of (A11.2) the integration with respect to μ yields a constant. Hence,

$$p(y_u|\mathbf{y}) \propto \exp[-\frac{1}{2(V(\mu)+\sigma^2)}(y_u-E(\mu))^2]. \tag{262.6}$$

According to (A11.1) and (A11.3) the predictive distribution for y_u is the normal distribution

$$y_u|\mathbf{y} \sim N(E(\mu),\ V(\mu)+\sigma^2) \tag{262.7}$$

with the expected value $E(\mu)$ from (211.18) and the variance $V(\mu)+\sigma^2$ with $V(\mu)$ from

(211.19). The expected value of the predicted observation y_u therefore agrees with the expected value of the parameter μ, while σ^2 is added to its variance, to obtain the variance of y_u.

We now substitute $\sigma_\mu^2 \to \infty$, which introduces a noninformative prior for the unknown parameter μ instead of the prior density (211.8). This substitution was already applied in Example 3 of Section 222 with the results from (222.16)

$$E(\mu) = \hat{\mu} \quad \text{and} \quad V(\mu) = \sigma^2/n .$$

In the case of a noninformative prior for the parameter μ the predictive distribution for y_u therefore follows with

$$y_u|\mathbf{y} \sim N(\hat{\mu},\ \sigma^2(n+1)/n) . \qquad (262.8)$$

Δ

27 Numerical Techniques

271 Monte Carlo Integration

In many applications the posterior density function for the unknown parameters resulting from Bayes' theorem can be readily written down. This is demonstrated for different models in Chapter 3. For estimating the parameters, for establishing confidence regions or for testing hypotheses these density functions have to be integrated with respect to the parameters. Frequently, however, the integration cannot be solved analytically, so that numerical methods have to be applied.

Well-known methods exist for the numerical integration by quadrature. Special approaches well suited for the integrals resulting from Bayesian inference can be found, for instance, in Press (1989, p.74). However, these methods become very inefficient with the increase of the dimension of the parameter space Θ. Monte Carlo integration helps to overcome this deficiency. It is based on generating random numbers by which an integral is computed as an expected value of a function of a random variable.

Let $\mathbf{x}$ contain the values of a random vector, also denoted by $\mathbf{x}$, and let $p(\mathbf{x})$ be a function of $\mathbf{x}$. We want to compute the integral

$$I = \int_A p(\mathbf{x})d\mathbf{x} \quad \text{with} \quad \mathbf{x} \in A \tag{271.1}$$

with A being the domain of the integration. Let $u(\mathbf{x})$ be the density function of $\mathbf{x}$. We rewrite I by

$$I = \int_A (p(\mathbf{x})/u(\mathbf{x}))u(\mathbf{x})d\mathbf{x},$$

so that the integral I can be interpreted as the expected value of the function $p(\mathbf{x})/u(\mathbf{x})$ of the random vector $\mathbf{x}$

$$I = E(p(\mathbf{x})/u(\mathbf{x})). \tag{271.2}$$

If a sequence of independent and identically distributed random vectors with the density $u(\mathbf{x})$ on A is generated giving $\mathbf{x}_1, \mathbf{x}_2, \ldots, \mathbf{x}_m$, the expected value (271.2) is estimated by

$$\hat{I} = \frac{1}{m} \sum_{i=1}^{m} (p(\mathbf{x}_i)/u(\mathbf{x}_i))$$

and (Frühwirth and Regler 1983, p.139; Hammersley and Handscomb 1964; p.57; Rubinstein 1981, p.122)

$$\int_A p(\mathbf{x})d\mathbf{x} = \frac{1}{m} \sum_{i=1}^{m} (p(\mathbf{x}_i)/u(\mathbf{x}_i)). \tag{271.3}$$

This method of integration is called *importance sampling*, since the generated data points $\mathbf{x}_i$ are concentrated due to the distribution $u(\mathbf{x})$ in the important part of A.

The main problem of the Monte Carlo integration (271.3) is that of finding an appropriate density $u(\mathbf{x})$. It should closely approximate the integrand $p(\mathbf{x})$, so that the ratio $p(\mathbf{x})/u(\mathbf{x})$ is nearly constant. This requirement will be seldom fulfilled, especially if $p(\mathbf{x})$ is a marginal density function which itself has been determined by numerical integration, cf. (273.2).

The simplest solution to this problem is to spread out the generated data points $\mathbf{x}_i$ evenly, which means assuming $u(\mathbf{x})$ as being uniformly distributed. If the domain A in (271.1) can be bordered by parallels to the coordinate axes, we obtain with $\mathbf{x}=(x_l)$ and $l=\{1,\ldots,u\}$ the density function

$$u(\mathbf{x}) = \begin{cases} \prod_{l=1}^{u} 1/(b_l-a_l) & \text{for} \quad a_l \le x_l \le b_l \\ 0 & \text{for} \quad x_l < a_l \ \text{and} \ x_l > b_l . \end{cases} \tag{271.4}$$

By substituting this result in (271.3) the integral (271.1) is computed by

$$\int_A p(\mathbf{x})d\mathbf{x} = [\prod_{l=1}^{u} (b_l-a_l)] \frac{1}{m} \sum_{i=1}^{m} p(\mathbf{x}_i), \tag{271.5}$$

which is called the *sample-mean* or *crude Monte Carlo method.*

If the domain A of the integration cannot be defined by parallels to the coordinate axes, then the density function of the uniform distribution of $\mathbf{x}$ in A is given by

$$u(\mathbf{x}) = 1/V_A, \tag{271.6}$$

where V_A denotes the volume of A. To generate uniformly distributed data points $\mathbf{x}_i$ in A, we may start from generating in a rectangular space enclosing A and omit all vectors outside A. With m vectors $\mathbf{x}_i$ in A we obtain by substituting (271.6) in (271.3)

$$\int_A p(\mathbf{x})d\mathbf{x} = (V_A/m) \sum_{i=1}^{m} p(\mathbf{x}_i). \tag{271.7}$$

All integrals which are encountered when computing estimates and confidence regions or when testing hypotheses can be solved by (271.5) or (271.7). This will be demonstrated in the next sections.

272 Computation of Estimates, Confidence Regions and Posterior Probabilities of Hypotheses

The Bayes estimate $\hat{\boldsymbol{\theta}}_B$ of the parameter vector $\boldsymbol{\theta}$ with $\boldsymbol{\theta}=(\theta_l)$ and $l\in\{1,\ldots,u\}$ is computed by the integral (231.6), whose domain is the parameter space Θ. Thus, the intervals $[a_l,b_l]$ with $l\in\{1,\ldots,u\}$ of (271.4) on the coordinate axes for $\boldsymbol{\theta}$ have to be chosen such that outside the region defined by the intervals the relation

$$\theta_l p(\boldsymbol{\theta}|\mathbf{y}) \le \varepsilon \tag{272.1}$$

is fulfilled, where ε denotes a small number. It is determined by the product of the parameter value and the density value which ceases to contribute to the integral.

Bayes' theorem (211.1) suggests working with posterior distributions which are not normalized and which shall be denoted by $\bar{p}(\boldsymbol{\theta}|\mathbf{y})$. The normalization constant follows from (211.5). The Bayes estimate $\hat{\boldsymbol{\theta}}_B$ of the parameter vector $\boldsymbol{\theta}$ is therefore computed instead of (231.6) by

$$\hat{\boldsymbol{\theta}}_B = \int_\Theta \boldsymbol{\theta}\, \bar{p}(\boldsymbol{\theta}|\mathbf{y})d\boldsymbol{\theta} / \int_\Theta \bar{p}(\boldsymbol{\theta}|\mathbf{y})d\boldsymbol{\theta}. \tag{272.2}$$

If we apply the Monte Carlo integration (271.5), we find

$$\hat{\boldsymbol{\theta}}_B = [\prod_{l=1}^{u}(b_l-a_l)]\,\frac{1}{m}\sum_{i=1}^{m}\boldsymbol{\theta}_i\bar{p}(\boldsymbol{\theta}_i|\mathbf{y})/[\prod_{l=1}^{u}(b_l-a_l)]\,\frac{1}{m}\sum_{i=1}^{m}\bar{p}(\boldsymbol{\theta}_i|\mathbf{y})$$

or

$$\hat{\boldsymbol{\theta}}_B = \sum_{i=1}^{m}\boldsymbol{\theta}_i\bar{p}(\boldsymbol{\theta}_i|\mathbf{y})/\sum_{i=1}^{m}\bar{p}(\boldsymbol{\theta}_i|\mathbf{y}), \tag{272.3}$$

where $\boldsymbol{\theta}_i$ denotes the random vector generated with a uniform distribution in the parameter space Θ bounded by the intervals resulting from (272.1).

The sum $\sum_{i=1}^{m}\bar{p}(\boldsymbol{\theta}_i|\mathbf{y})$ in (272.3) represents the normalization factor. The normalized posterior density value $p(\boldsymbol{\theta}_i|\mathbf{y})$ of the generated random vector $\boldsymbol{\theta}_i$ therefore follows with

$$p(\boldsymbol{\theta}_i|\mathbf{y}) = \bar{p}(\boldsymbol{\theta}_i|\mathbf{y})/\sum_{i=1}^{m}\bar{p}(\boldsymbol{\theta}_i|\mathbf{y}). \tag{272.4}$$

The values $p(\boldsymbol{\theta}_i|\mathbf{y})$ can be interpreted as the values of a density function of a discrete random vector $\boldsymbol{\theta}$ with the values $\boldsymbol{\theta}_i$ for $i\in\{1,\ldots,m\}$ originating from the random number generator. The density function $p(\boldsymbol{\theta}_i|\mathbf{y})$ fulfills the conditions for the density function of a discrete random vector (Koch 1988a, p.98)

$$p(\boldsymbol{\theta}_i|\mathbf{y}) \geq 0 \quad \text{and} \quad \sum_{i=1}^{m} p(\boldsymbol{\theta}_i|\mathbf{y}) = 1. \tag{272.5}$$

By means of this density function we may compute the confidence region for the parameter vector $\boldsymbol{\theta}$ defined by (241.2). Let the density values $p(\boldsymbol{\theta}_i|\mathbf{y})$ be ordered according to increasing values, so that the sequence $p(\boldsymbol{\theta}_j|\mathbf{y})$ with $j\in\{1,\ldots,m\}$ is obtained. Then the value b in (242.1) is determined by

$$b = p(\boldsymbol{\theta}_o|\mathbf{y}), \tag{272.6}$$

where $p(\boldsymbol{\theta}_o|\mathbf{y})$ follows from

$$\sum_{j=1}^{o} p(\boldsymbol{\theta}_j|\mathbf{y}) = \alpha \tag{272.7}$$

with α giving the content $1-\alpha$ of the confidence region. The equation (272.7) can only be approximately fulfilled. But the more vectors $\boldsymbol{\theta}_j$ are generated, the smaller will be the increase from $p(\boldsymbol{\theta}_j|\mathbf{y})$ to $p(\boldsymbol{\theta}_{j+1}|\mathbf{y})$ and the better the approximation. Already a linear interpolation between the density values $p(\boldsymbol{\theta}_j|\mathbf{y})$ and $p(\boldsymbol{\theta}_{j+1}|\mathbf{y})$ will improve the approximation. When adding additional density values by interpolation, a new normalization factor has to be computed.

The boundary of the confidence region is determined according to (242.1) by the vectors $\boldsymbol{\theta}_b$ for which

$$b = p(\boldsymbol{\theta}_b|\mathbf{y})$$

is fulfilled. For a determination of the vectors $\boldsymbol{\theta}_b$ at the boundary, we visualize the generated vectors $\boldsymbol{\theta}_i$ as points of the parameter space Θ and select neighboring points $\boldsymbol{\theta}_i$ and $\boldsymbol{\theta}_j$ with

$$p(\boldsymbol{\theta}_i|\mathbf{y}) \leq b \quad \text{and} \quad p(\boldsymbol{\theta}_j|\mathbf{y}) \geq b. \tag{272.8}$$

By interpolation between these points the points $\boldsymbol{\theta}_b$ at the boundary are obtained.

For computing posterior probabilities associated with hypotheses the integrals (253.2) have to be solved. Again we assume that we work with a posterior density function $\bar{p}(\boldsymbol{\theta}|\mathbf{y})$ which is not normalized. Thus, we obtain with (211.5) instead of (253.2)

$$P(H_k|\mathbf{y}) = \int_{\Theta_k} \bar{p}(\boldsymbol{\theta}|\mathbf{y})d\boldsymbol{\theta} / \int_{\Theta} \bar{p}(\boldsymbol{\theta}|\mathbf{y})d\boldsymbol{\theta} \quad \text{for} \quad k\in\{0,1\}. \tag{272.9}$$

We generate random vectors $\boldsymbol{\theta}_i$ for $i\in\{1,\ldots,m\}$ with uniform distributions in the parameter space Θ bounded by the intervals $[a_l,b_l]$ on the coordinate axes for $\boldsymbol{\theta}$. Correspondingly to (272.1) the intervals are established such that outside the region, defined by the intervals, the density values cease to contribute to the integrals. In addition we generate random vectors $\boldsymbol{\theta}_{jk}$ for $j\in\{1,\ldots,n\}$ and $k\in\{0,1\}$ with uniform distributions in the subspace Θ_k for $k\in\{0,1\}$. Let these subspaces be defined by the intervals $[a_{lk}, b_{lk}]$ with $l\in\{1,\ldots,u\}$. Hence, the Monte Carlo integration of (272.9) follows with (271.5) by

$$P(H_k|\mathbf{y}) = [\prod_{l=1}^{u}(b_{lk}-a_{lk})]\ \frac{1}{n}\sum_{j=1}^{n}\bar{p}(\boldsymbol{\theta}_{jk}|\mathbf{y})/$$

$$[\prod_{l=1}^{u}(b_l-a_l)]\ \frac{1}{m}\sum_{i=1}^{m}\bar{p}(\boldsymbol{\theta}_i|\mathbf{y}) \quad \text{for} \quad k\in\{0,1\}. \tag{272.10}$$

The posterior odds (253.6) are thus computed by

$$P(H_o|\mathbf{y})/P(H_1|\mathbf{y}) = [\prod_{l=1}^{u}(b_{lo}-a_{lo})]\ \sum_{j=1}^{n}\bar{p}(\boldsymbol{\theta}_{jo}|\mathbf{y})/$$

$$[\prod_{l=1}^{u}(b_{l1}-a_{l1})]\ \sum_{j=1}^{n}\bar{p}(\boldsymbol{\theta}_{j1}|\mathbf{y}). \tag{272.11}$$

If the subspaces Θ_o and Θ_1 cannot be defined by intervals, but have volumes V_o and V_1, respectively, we obtain with (271.7) instead of (272.11)

$$P(H_o|\mathbf{y})/P(H_1|\mathbf{y}) = V_o\sum_{j=1}^{n}\bar{p}(\boldsymbol{\theta}_{jo}|\mathbf{y})/V_1\sum_{j=1}^{n}\bar{p}(\boldsymbol{\theta}_{j1}|\mathbf{y}). \tag{272.12}$$

Example 1: Examples for the computation of the Bayes estimate according to (272.3) and the confidence limits according to (272.6) are given in the Examples 1 of Section 322 and 342. Δ

273 Marginal Distributions and Transformation of Variables

Frequently the distribution of a subset of the set of parameters $\boldsymbol{\theta}$ is needed. It is obtained as a marginal distribution (Koch 1988a, p.105) of the posterior density function $p(\boldsymbol{\theta}|\mathbf{y})$ resulting from Bayes' theorem (211.1). With $\boldsymbol{\theta}=|\boldsymbol{\theta}_1', \boldsymbol{\theta}_2'|'$, where $\boldsymbol{\theta}_1$ denotes the parameter vector whose marginal distribution is needed, we obtain the marginal posterior density function $p(\boldsymbol{\theta}_1|\mathbf{y})$ of $\boldsymbol{\theta}_1$ by

$$p(\boldsymbol{\theta}_1|\mathbf{y}) = \int_{\Theta_2} p(\boldsymbol{\theta}_1, \boldsymbol{\theta}_2|\mathbf{y})d\boldsymbol{\theta}_2. \tag{273.1}$$

The parameter space for $\boldsymbol{\theta}_2$ is denoted by Θ_2 with $\Theta_2\subset\Theta$. The marginal distribution $p(\boldsymbol{\theta}_1|\mathbf{y})$ shall be determined by a Monte Carlo integration and it shall be used again in an ensuing Monte Carlo integration for computing, for instance, the confidence region for the parameter vector $\boldsymbol{\theta}_1$.

Again we will work with a posterior density function $\bar{p}(\boldsymbol{\theta}_1, \boldsymbol{\theta}_2 | \mathbf{y})$ which is not normalized. Thus, we obtain with (271.5) instead of (273.1)

$$\bar{p}(\boldsymbol{\theta}_{1i} | \mathbf{y}) = \sum_{j=1}^{o} \bar{p}(\boldsymbol{\theta}_{1i}, \boldsymbol{\theta}_{2j} | \mathbf{y}), \tag{273.2}$$

where the vectors $\boldsymbol{\theta}_{2j}$ for $j \in \{1, \ldots, o\}$ are generated with uniform distributions in the space Θ_2 for $\boldsymbol{\theta}_2$. Since non-normalized density functions are used, the constants in (271.5) may be omitted. The vectors $\boldsymbol{\theta}_{1i}$ result from the generation of random vectors with uniform distributions for the Monte Carlo integration in connection with the marginal posterior density $\bar{p}(\boldsymbol{\theta}_1 | \mathbf{y})$, for instance when computing the estimate $\hat{\boldsymbol{\theta}}_{1B}$ of $\boldsymbol{\theta}_1$ according to (272.3), when establishing the confidence region for $\boldsymbol{\theta}_1$ according to (272.6) or when testing hypotheses for $\boldsymbol{\theta}_1$ according to (272.11).

Occasionally we have to go one step further, if the parameter vector $\boldsymbol{\theta}_1$ in (273.1) needs to be transformed and the distribution of a subset of the transformed parameters has to be computed.

Let the parameter vector $\boldsymbol{\theta}_1$, which shall be of dimension q×1, be transformed to the q×1 parameter vector $\boldsymbol{\gamma}_1$. Let the inverse transformation with $\boldsymbol{\theta}_1 = (\theta_{1i})$ and $\boldsymbol{\gamma}_1 = (\gamma_{1i})$ be given by

$$\boldsymbol{\theta}_1 = \mathbf{g}(\boldsymbol{\gamma}_1) \quad \text{or} \quad \theta_{1i} = g_i(\gamma_{11}, \ldots, \gamma_{1q}) \quad \text{for} \quad i \in \{1, \ldots, q\}. \tag{273.3}$$

If $p(\boldsymbol{\gamma}_1 | \mathbf{y})$ denotes the posterior density of the transformed parameters $\boldsymbol{\gamma}$, we obtain (Koch 1988a, p.108)

$$p(\boldsymbol{\gamma}_1 | \mathbf{y}) = p(\mathbf{g}(\boldsymbol{\gamma}_1) | \mathbf{y}) |\det \mathbf{J}|, \tag{273.4}$$

where $|\det \mathbf{J}|$ denotes the absolute value of the Jacobian $\det \mathbf{J}$ with

$$\mathbf{J} = (\partial g_i / \partial \gamma_{1j}) \quad \text{for} \quad i, j \in \{1, \ldots, q\}. \tag{273.5}$$

The vector $\boldsymbol{\gamma}_1$ is again partitioned into $\boldsymbol{\gamma}_1 = |\boldsymbol{\gamma}_{11}', \boldsymbol{\gamma}_{12}'|'$. Only the marginal posterior density function $p(\boldsymbol{\gamma}_{11} | \mathbf{y})$ of the parameter vector $\boldsymbol{\gamma}_{11}$ is needed. Hence,

$$p(\boldsymbol{\gamma}_{11} | \mathbf{y}) = \int_{\Gamma_{12}} p(\boldsymbol{\gamma}_{11}, \boldsymbol{\gamma}_{12} | \mathbf{y}) d\boldsymbol{\gamma}_{12}, \tag{273.6}$$

where Γ_{12} denotes the parameter space for $\boldsymbol{\gamma}_{12}$. By substituting (273.4) we find

$$p(\boldsymbol{\gamma}_{11} | \mathbf{y}) = \int_{\Gamma_{12}} p[\mathbf{g}(\boldsymbol{\gamma}_{11}, \boldsymbol{\gamma}_{12}) | \mathbf{y}] |\det \mathbf{J}| d\boldsymbol{\gamma}_{12}$$

and by substituting (273.1)

$$p(\boldsymbol{\gamma}_{11} | \mathbf{y}) = \int_{\Gamma_{12}} \int_{\Theta_2} p[\mathbf{g}(\boldsymbol{\gamma}_{11}, \boldsymbol{\gamma}_{12}), \boldsymbol{\theta}_2 | \mathbf{y}] |\det \mathbf{J}| d\boldsymbol{\theta}_2 \, d\boldsymbol{\gamma}_{12}. \tag{273.7}$$

The marginal posterior density $p(\gamma_{11}|y)$ shall be computed by a Monte Carlo integration, in order to be used again in a further Monte Carlo integration. Correspondingly to the step from (273.1) to (273.2) we obtain

$$\bar{p}(\gamma_{11i}|y) = \sum_{j=1}^{o}\sum_{k=1}^{r} \bar{p}[g(\gamma_{11i},\gamma_{12j}),\theta_{2k}|y]\,|\det J|. \qquad (273.8)$$

The vector γ_{11i} stems from the Monte Carlo integration in connection with the non-normalized density $\bar{p}(\gamma_{11}|y)$, the vector γ_{12j} from the generation of the vectors with uniform distributions in the space Γ_{12} for γ_{12} and the vector θ_{2k} from the generation with uniform distributions in Θ_2 for θ_2.

With the density values $\bar{p}(\gamma_{11i}|y)$ from (273.8) we may compute the estimate $\hat{\gamma}_{11B}$ of γ_{11} by (272.3), establish the confidence region for γ_{11} by (272.6) or test hypotheses for γ_{11} by (272.11).

Example 1: Examples for computing the marginal density function from (273.2) are given in the Examples 1 of Section 322 and 342. An example for applying (273.8) can be found in Koch (1988c). Δ

274 Approximate Computation of Marginal Distributions

If the vector θ_2 in (273.1) or the vectors γ_{12} and θ_2 in (273.7) contain a large number of unknown parameters, the Monte Carlo integration (273.2) or (273.8) is very time-consuming even with fast electronic computers. An approximate computation of the marginal distribution instead of a numerical integration shall therefore be developed. The method will be demonstrated for the definition (273.1) of the marginal distribution of θ_1, but it may, of course, be applied to (273.7).

We have

$$p(\theta_1|y) = \int_{\Theta_2} p(\theta_1,\theta_2|y)d\theta_2 = p(\theta_1,\theta_{2c}|y), \qquad (274.1)$$

if θ_{2c} denotes an appropriately chosen constant. We will check now, whether θ_{2c} may be replaced by an estimate $\hat{\theta}_2$ of θ_2, thus,

$$\theta_{2c} = \hat{\theta}_2. \qquad (274.2)$$

The posterior density function of the parameter vectors θ_1 and θ_2 is expressed with (261.1) by

$$p(\theta_1,\theta_2|y) = p(\theta_1|\theta_2,y)p(\theta_2|y), \qquad (274.3)$$

where $p(\boldsymbol{\theta}_1|\boldsymbol{\theta}_2,\mathbf{y})$ denotes the conditional posterior density function of $\boldsymbol{\theta}_1$ given $\boldsymbol{\theta}_2$ and $\mathbf{y}$ and $p(\boldsymbol{\theta}_2|\mathbf{y})$ the marginal posterior density function of $\boldsymbol{\theta}_2$. Corresponding to (273.1) the marginal posterior distribution $p(\boldsymbol{\theta}_1|\mathbf{y})$ of $\boldsymbol{\theta}_1$ is obtained from (274.3) by

$$p(\boldsymbol{\theta}_1|\mathbf{y}) = \int_{\Theta_2} p(\boldsymbol{\theta}_1|\boldsymbol{\theta}_2,\mathbf{y})p(\boldsymbol{\theta}_2|\mathbf{y})d\boldsymbol{\theta}_2 = p(\boldsymbol{\theta}_1|\boldsymbol{\theta}_{2k},\mathbf{y}), \tag{274.4}$$

where $\boldsymbol{\theta}_{2k}$ denotes an appropriately chosen constant.

The marginal posterior distribution $p(\boldsymbol{\theta}_2|\mathbf{y})$ of $\boldsymbol{\theta}_2$ contains the information on $\boldsymbol{\theta}_2$ resulting from the observations $\mathbf{y}$ and the prior information. If the density function $p(\boldsymbol{\theta}_2|\mathbf{y})$ has a pointed shape, its probability mass is concentrated over a small region in the vicinity of its modal value $\bar{\boldsymbol{\theta}}_2$, i.e. the value for which $p(\boldsymbol{\theta}_2|\mathbf{y})$ attains a maximum. Then $p(\boldsymbol{\theta}_2|\mathbf{y})$ contains so much information on $\boldsymbol{\theta}_2$ that values for $\boldsymbol{\theta}_2$, which are not close to $\bar{\boldsymbol{\theta}}_2$, may be excluded. Hence, the integration with respect to $\boldsymbol{\theta}_2$ is approximately equivalent to choosing for the constant $\boldsymbol{\theta}_{2k}$ the modal value $\bar{\boldsymbol{\theta}}_2$

$$\boldsymbol{\theta}_{2k} = \bar{\boldsymbol{\theta}}_2 . \tag{274.5}$$

According to (232.4) $\bar{\boldsymbol{\theta}}_2$ is the generalized maximum likelihood estimate of $\boldsymbol{\theta}_2$.

On the other hand, if $p(\boldsymbol{\theta}_2|\mathbf{y})$ has a flat shape, there is not much information on $\boldsymbol{\theta}_2$ available. In order to justify under these circumstances a choice according (274.5), additional information on $\boldsymbol{\theta}_2$ by the data $\mathbf{y}$ or by prior information has to be collected.

With (274.5) we obtain instead of (274.4)

$$p(\boldsymbol{\theta}_1|\mathbf{y}) = p(\boldsymbol{\theta}_1|\bar{\boldsymbol{\theta}}_2,\mathbf{y})$$

and from (274.3)

$$p(\boldsymbol{\theta}_1|\bar{\boldsymbol{\theta}}_2,\mathbf{y}) \propto p(\boldsymbol{\theta}_1,\bar{\boldsymbol{\theta}}_2|\mathbf{y}) .$$

The marginal posterior density function $p(\boldsymbol{\theta}_1|\mathbf{y})$ then follows with

$$p(\boldsymbol{\theta}_1|\mathbf{y}) \propto p(\boldsymbol{\theta}_1,\bar{\boldsymbol{\theta}}_2|\mathbf{y}) . \tag{274.6}$$

Thus, $p(\boldsymbol{\theta}_1|\mathbf{y})$ can be approximately determined by the maximum likelihood estimate $\bar{\boldsymbol{\theta}}_2$ of $\boldsymbol{\theta}_2$, if the density function $p(\boldsymbol{\theta}_2|\mathbf{y})$ has a pointed shape. In such a case the modal value lies in the vicinity of the expected value, if the two values do not coincide. The maximum likelihood estimate $\bar{\boldsymbol{\theta}}_2$ of $\boldsymbol{\theta}_2$ in (274.6) may therefore be replaced by the Bayes estimate $\hat{\boldsymbol{\theta}}_{2B}$ of $\boldsymbol{\theta}_2$ from (231.5) so that we obtain

$$p(\boldsymbol{\theta}_1|\mathbf{y}) \propto p(\boldsymbol{\theta}_1,\hat{\boldsymbol{\theta}}_{2B}|\mathbf{y}) . \tag{274.7}$$

The density function $p(\boldsymbol{\theta}_2|\mathbf{y})$ has a pointed shape, if the observations $\mathbf{y}$ contain accurate information on the parameter vector $\boldsymbol{\theta}_2$ or if the prior information on $\boldsymbol{\theta}_2$ is accurate.

Example 1: Examples for an approximate computation of a marginal posterior density function from (274.6) is given in Example 1 of Section 322 and in Example 1 of Section 365. Additional examples can be found in Koch (1989). Applications of (274.7) are contained in Section 373. Δ

3 Models and Special Applications

When observations are taken which contain information on unknown parameters, the relation between the unknown parameters and the observations has to be defined. In addition, the statistical properties of the observations need to be defined. These definitions establish the *model*. It serves to analyze the data for the statistical inference on the parameters.

In general the expected values of the observations are defined as given functions of the unknown parameters, and the covariance matrix of the observations is assumed as known except for an unknown factor. Due to the central limit theorem (Cramer 1946, p.214,316), we may assume the observations as being normally distributed. Since the density function of a normally distributed random vector is according to (A21.3) uniquely specified by its expected value and its covariance matrix, the assumptions for the model determine the likelihood function. Thus, with the likelihood function determined by the model and the prior distribution chosen according to the available information, Bayes′ theorem (211.1) readily leads to the posterior density function of the unknown parameters.

In addition to the statistical inference on unknown parameters, special applications are presented in this chapter such as the inference for unknown parameters based on special likelihood functions, which originate from robust estimation. The problem of the classification and the reconstruction of digital images is also discussed.

31 Linear Models

311 Definition and Likelihood Function

The most important model of the standard statistical techniques is the *Gauss-Markoff model* or the *linear model* (Koch 1988a, p.182). An estimation of the parameters of this model is also called a *regression analysis*. To obtain for the Bayesian analysis a model which corresponds to the Gauss-Markoff model, one has to be aware that the parameters of the Gauss-Markoff model are fixed quantities, while in the Bayesian analysis the parameters are random variables. Hence, the expected values of the observations and the covariance matrix have to be defined under the condition that the unknown parameters take on fixed values.

Definition: Let $\mathbf{X}$ be an n×u matrix of given coefficients with full column rank, i.e. $\operatorname{rank}\mathbf{X}=u$, $\boldsymbol{\beta}$ a u×1 vector of unknown random parameters, $\mathbf{y}$ an n×1 random vector of observations and σ^2 an unknown random parameter, which is called *variance factor* or *variance of unit weight*. Let $E(\mathbf{y}|\boldsymbol{\beta})$ and $D(\mathbf{y}|\sigma^2)$ be the expected value and the covariance matrix of the observation vector $\mathbf{y}$ under the condition that the unknown parameters $\boldsymbol{\beta}$ and σ^2 are given, then

$$E(\mathbf{y}|\boldsymbol{\beta}) = \mathbf{X}\boldsymbol{\beta} \quad \text{with} \quad D(\mathbf{y}|\sigma^2) = \sigma^2\mathbf{I}$$

is called a *linear model*. (311.1)

An equivalent formulation of this model is given by the so-called observation equations

$$\mathbf{y}+\mathbf{e} = \mathbf{X}\boldsymbol{\beta} \quad \text{with} \quad E(\mathbf{e}|\boldsymbol{\beta}) = \mathbf{0} \quad \text{and} \quad D(\mathbf{e}|\sigma^2) = D(\mathbf{y}|\sigma^2), \tag{311.2}$$

where $\mathbf{e}$ denotes the n×1 random vector of the errors of the observations.

The model (311.1) or (311.2) is of sufficient generality. The more general model

$$E(\bar{\mathbf{y}}|\boldsymbol{\beta}) = \bar{\mathbf{X}}\boldsymbol{\beta} \quad \text{or} \quad \bar{\mathbf{y}} + \bar{\mathbf{e}} = \bar{\mathbf{X}}\boldsymbol{\beta} \quad \text{with} \quad D(\bar{\mathbf{y}}|\sigma^2) = \sigma^2\mathbf{P}^{-1}, \; E(\bar{\mathbf{e}}|\boldsymbol{\beta}) = \mathbf{0},$$
$$D(\bar{\mathbf{e}}|\sigma^2) = D(\bar{\mathbf{y}}|\sigma^2), \tag{311.3}$$

where the n×n positive definite matrix $\mathbf{P}$ is called the *weight matrix*, can be transformed to (311.1) or (311.2) with

$$\mathbf{P} = \mathbf{G}\mathbf{G}', \; \mathbf{X} = \mathbf{G}'\bar{\mathbf{X}}, \; \mathbf{y} = \mathbf{G}'\bar{\mathbf{y}}, \; \mathbf{e} = \mathbf{G}'\bar{\mathbf{e}} \tag{311.4}$$

(Koch 1988a, p.183). The matrix $\mathbf{G}$ results from the Cholesky factorization of $\mathbf{P}$ (Koch 1988a, p.36) and denotes a regular lower triangular matrix.

As mentioned in Chapter 3, we will assume the normal distribution for the observation vector $\mathbf{y}$ under the condition that $\boldsymbol{\beta}$ and σ^2 are given

$$\mathbf{y}|\boldsymbol{\beta},\sigma^2 \sim N(\mathbf{X}\boldsymbol{\beta},\sigma^2\mathbf{I}). \tag{311.5}$$

The likelihood function, i.e. the density function of the observations $\mathbf{y}$ given the parameters $\boldsymbol{\beta}$ and σ^2, then follows with (A21.1) by

$$p(\mathbf{y}|\boldsymbol{\beta},\sigma^2) = (2\pi)^{-n/2}(\det\sigma^2\mathbf{I})^{-1/2}\exp[-\frac{1}{2\sigma^2}(\mathbf{y}-\mathbf{X}\boldsymbol{\beta})'(\mathbf{y}-\mathbf{X}\boldsymbol{\beta})]$$

or

$$p(\mathbf{y}|\boldsymbol{\beta},\sigma^2) = (2\pi\sigma^2)^{-n/2}\exp[-\frac{1}{2\sigma^2}(\mathbf{y}-\mathbf{X}\boldsymbol{\beta})'(\mathbf{y}-\mathbf{X}\boldsymbol{\beta})]. \tag{311.6}$$

If the weight or precision parameter τ is used in (311.1) with

$$\tau = 1/\sigma^2, \tag{311.7}$$

we obtain instead of (311.5) the distribution

$$\mathbf{y}|\boldsymbol{\beta},\tau \sim N(\mathbf{X}\boldsymbol{\beta},\tau^{-1}\mathbf{I}). \tag{311.8}$$

The likelihood function therefore follows with

$$p(\mathbf{y}|\boldsymbol{\beta},\tau) = (2\pi)^{-n/2}\tau^{n/2}\exp[-\frac{\tau}{2}(\mathbf{y}-\mathbf{X}\boldsymbol{\beta})'(\mathbf{y}-\mathbf{X}\boldsymbol{\beta})] \tag{311.9}$$

and by means of the sufficient statistics (224.4) because of (224.7) with

$$p(\mathbf{y}|\boldsymbol{\beta},\tau) = (2\pi)^{-n/2}\tau^{n/2}\exp\{-\frac{\tau}{2}[(n-u)\hat{\sigma}^2 + (\boldsymbol{\beta}-\hat{\boldsymbol{\beta}})'\mathbf{X}'\mathbf{X}(\boldsymbol{\beta}-\hat{\boldsymbol{\beta}})]\}. \tag{311.10}$$

Example 1: Let the length μ of a straight line be measured n times and let the n observations y_i be collected in the nx1 random vector $\mathbf{y}$ with $\mathbf{y}=(y_i)$. Let the observations y_i be independent, have the expectation $E(y_i|\mu)=\mu$ and variance $V(y_i|\sigma^2)=\sigma^2$. Let μ and σ^2 be the unknown parameters and e_i the errors. We therefore have the linear model

$$\begin{matrix} y_1 + e_1 = \mu \\ y_2 + e_2 = \mu \\ \cdots\cdots\cdots \\ y_n + e_n = \mu \end{matrix} \quad \text{and} \quad D(\begin{bmatrix} y_1 \\ y_2 \\ \cdots \\ y_n \end{bmatrix}|\sigma^2) = \sigma^2\begin{bmatrix} 1 & 0 & \dots & 0 \\ 0 & 1 & \dots & 0 \\ \cdots & \cdots & \cdots & \cdots \\ 0 & 0 & \dots & 1 \end{bmatrix} \tag{311.11}$$

or in the representation (311.1)

$$E(\mathbf{y}|\boldsymbol{\beta}) = \mathbf{X}\boldsymbol{\beta} \quad \text{with} \quad D(\mathbf{y}|\sigma^2) = \sigma^2\mathbf{I} \tag{311.12}$$

with

$$\mathbf{X} = [1,1,\dots,1]' \quad \text{and} \quad \boldsymbol{\beta} = \mu.$$

Let in addition the observations y_i be normally distributed, then with (311.5)

$$\mathbf{y}|\mu,\sigma^2 \sim N(\mathbf{X}\boldsymbol{\beta},\sigma^2\mathbf{I}). \tag{311.13}$$

From (311.11) and (311.12) we have

$$(\mathbf{y}-\mathbf{X}\boldsymbol{\beta})'(\mathbf{y}-\mathbf{X}\boldsymbol{\beta}) = \sum_{i=1}^{n}(y_i-\mu)^2,$$

so that the likelihood function follows with (311.6) by

$$p(\mathbf{y}|\mu,\sigma^2) = (2\pi\sigma^2)^{-n/2}\exp[-\frac{1}{2\sigma^2}\sum_{i=1}^{n}(y_i-\mu)^2]. \tag{311.14}$$

The same likelihood function was already derived with (224.21) for the Example 1 of Section 224. Δ

312 Noninformative Priors

We will assume a noninformative prior density function for the unknown parameters $\boldsymbol{\beta}$ and σ^2 of the linear model (311.1). We apply Jeffrey's invariance principle to derive the priors and consider $\boldsymbol{\beta}$ and σ^2 as being independent. Hence, we obtain with (222.11) and (311.7) after omitting the constants in (311.9)

$$\ln p(\mathbf{y}|\boldsymbol{\beta},\tau) \propto \frac{n}{2}\ln\tau - \tau(\mathbf{y}-\mathbf{X}\boldsymbol{\beta})'(\mathbf{y}-\mathbf{X}\boldsymbol{\beta})/2. \tag{312.1}$$

The derivative with respect to $\boldsymbol{\beta}$ follows with

$$\partial\ln p(\mathbf{y}|\boldsymbol{\beta},\tau)/\partial\boldsymbol{\beta} \propto -\tau(-2\mathbf{X}'\mathbf{y}+2\mathbf{X}'\mathbf{X}\boldsymbol{\beta})/2.$$

The second derivative with respect to $\boldsymbol{\beta}$ gives constants. The noninformative prior density function $p(\boldsymbol{\beta})$ for the parameter vector $\boldsymbol{\beta}$ therefore follows with (222.10) by

$$p(\boldsymbol{\beta}) \propto \text{const}. \tag{312.2}$$

The derivative of (312.1) with respect to τ gives

$$\partial\ln p(\mathbf{y}|\boldsymbol{\beta},\tau)/\partial\tau \propto \frac{n}{2\tau} - \frac{1}{2}(\mathbf{y}-\mathbf{X}\boldsymbol{\beta})'(\mathbf{y}-\mathbf{X}\boldsymbol{\beta})$$

and the second derivative

$$\partial^2\ln p(\mathbf{y}|\boldsymbol{\beta},\tau)/\partial\tau^2 \propto -\frac{n}{2\tau^2}. \tag{312.3}$$

Thus, we find with (222.10) the noninformative prior density function $p(\tau)$ for the parameter τ by

$$p(\tau) \propto 1/\tau. \tag{312.4}$$

Applying Bayes' theorem (211.1) the posterior density function of the parameters $\boldsymbol{\beta}$ and τ then follows with (311.10), (312.2) and (312.4) after omitting the constants by

$$p(\boldsymbol{\beta},\tau|\mathbf{y}) \propto \tau^{n/2-1} \exp\{-\frac{\tau}{2}[(n-u)\hat{\sigma}^2 + (\boldsymbol{\beta}-\hat{\boldsymbol{\beta}})'\mathbf{X}'\mathbf{X}(\boldsymbol{\beta}-\hat{\boldsymbol{\beta}})]\}. \tag{312.5}$$

By comparing this density function with (A23.1) it becomes evident because of $n/2-1=u/2+(n-u)/2-1$ that (312.5) is the density of the normal-gamma distribution. Thus,

$$\boldsymbol{\beta},\tau|\mathbf{y} \sim NG(\hat{\boldsymbol{\beta}},(\mathbf{X}'\mathbf{X})^{-1},(n-u)\hat{\sigma}^2/2,(n-u)/2). \tag{312.6}$$

The marginal posterior distribution of $\boldsymbol{\beta}$ is according to (A23.3) the multivariate t-distribution

$$\boldsymbol{\beta}|\mathbf{y} \sim t(\hat{\boldsymbol{\beta}},\hat{\sigma}^2(\mathbf{X}'\mathbf{X})^{-1},n-u) \tag{312.7}$$

with the expected value from (224.4) and (A22.7)

$$E(\boldsymbol{\beta}) = \hat{\boldsymbol{\beta}} = (\mathbf{X}'\mathbf{X})^{-1}\mathbf{X}'\mathbf{y} \tag{312.8}$$

and the covariance matrix

$$D(\boldsymbol{\beta}) = \frac{n-u}{n-u-2}\hat{\sigma}^2(\mathbf{X}'\mathbf{X})^{-1} \tag{312.9}$$

with

$$\hat{\sigma}^2 = \frac{1}{n-u}(\mathbf{y}-\mathbf{X}\hat{\boldsymbol{\beta}})'(\mathbf{y}-\mathbf{X}\hat{\boldsymbol{\beta}}). \tag{312.10}$$

Applying (231.5) and (231.7) the Bayes estimate $\hat{\boldsymbol{\beta}}_B$ of $\boldsymbol{\beta}$ is obtained by

$$\hat{\boldsymbol{\beta}}_B = \hat{\boldsymbol{\beta}} = (\mathbf{X}'\mathbf{X})^{-1}\mathbf{X}'\mathbf{y} \tag{312.11}$$

and the covariance matrix $\boldsymbol{\Sigma}_{\hat{\beta}}$ of the Bayes estimate $\hat{\boldsymbol{\beta}}_B$ by

$$\boldsymbol{\Sigma}_{\hat{\beta}} = \frac{n-u}{n-u-2}\hat{\sigma}^2(\mathbf{X}'\mathbf{X})^{-1}. \tag{312.12}$$

The estimate (312.11) is in addition the best linear unbiased estimate of the parameter vector $\boldsymbol{\beta}$ of the Gauss-Markoff model by the standard statistical techniques (Koch 1988a, p.187). It is identical with the estimate of the method of least squares and for normally distributed observations with the maximum likelihood estimate (Koch 1988a, p.191). Hence, in the case of the noninformative priors (312.2) and (312.4), the Bayes estimate (312.11) of the unknown parameters $\boldsymbol{\beta}$ is identical with the estimates of the standard statistical techniques. The covariance matrix $\boldsymbol{\Sigma}_{\hat{\beta}}$ of the estimate differs by the factor $(n-u)/(n-u-2)$ (Koch 1988a, p.193).

Because of (312.6) the marginal posterior distribution of the weight parameter τ is according to (A23.4) the gamma distribution

$$\tau|\mathbf{y} \sim G((n-u)\hat{\sigma}^2/2,(n-u)/2). \tag{312.13}$$

The distribution of the variance σ^2 of unit weight is therefore according to (A13.1) the inverted gamma distribution. Its expected value and variance (A13.2) give because of

(231.5) and (231.7) the Bayes estimate $\hat{\sigma}_B^2$ of σ^2 and its variance $V(\hat{\sigma}_B^2)$. Thus,

$$\hat{\sigma}_B^2 = \frac{n-u}{n-u-2}\hat{\sigma}^2 \tag{312.14}$$

with

$$V(\hat{\sigma}_B^2) = \frac{2(n-u)^2(\hat{\sigma}^2)^2}{(n-u-2)^2(n-u-4)} \ . \tag{312.15}$$

For the sake of comparison the best unbiased estimate of σ^2 resulting from the standard techniques is $\hat{\sigma}^2$ and its variance $2(\hat{\sigma}^2)^2/(n-u)$ (Koch 1988a, p.277).

Example 1: The Bayes estimates just derived shall be applied to the special linear model (311.11) of the Example 1 of Section 311. We obtain the Bayes estimate $\hat{\mu}_B$ of the unknown parameter μ from (312.11) and its variance $V(\hat{\mu}_B)$ from (312.12). Thus,

$$\hat{\mu}_B = \frac{1}{n}\sum_{i=1}^{n} y_i , \tag{312.16}$$

which is the well-known mean. As already mentioned, this estimate is identical with the estimate of the standard techniques. Furthermore with $\hat{\sigma}^2$ from (312.10)

$$\hat{\sigma}^2 = \frac{1}{n-1}\sum_{i=1}^{n} (y_i - \hat{\mu}_B)^2 \tag{312.17}$$

we find

$$V(\hat{\mu}_B) = \frac{(n-1)}{(n-3)}\frac{\hat{\sigma}^2}{n} \ . \tag{312.18}$$

As mentioned, the result of the standard techniques is $\hat{\sigma}^2/n$.

The Bayes estimate $\hat{\sigma}_B^2$ of the variance σ^2 of unit weight and the variance $V(\hat{\sigma}_B^2)$ of the estimate follow from (312.14) and (312.15) with

$$\hat{\sigma}_B^2 = \frac{n-1}{n-3}\hat{\sigma}^2 \tag{312.19}$$

and

$$V(\hat{\sigma}_B^2) = \frac{2(n-1)^2(\hat{\sigma}^2)^2}{(n-3)^2(n-5)} \ . \tag{312.20}$$

As already explained, the estimate of σ^2 by the standard statistical techniques is $\hat{\sigma}^2$ and its variance $2(\hat{\sigma}^2)^2/(n-1)$. Δ

Since the parameter vector $\boldsymbol{\beta}$ has the multivariate t-distribution (312.7), a confidence region for $\boldsymbol{\beta}$ is easily established. According to (A22.1) the posterior density function for $\boldsymbol{\beta}$ monotonically decreases, if the quadratic form $(\boldsymbol{\beta}-\hat{\boldsymbol{\beta}})'X'X(\boldsymbol{\beta}-\hat{\boldsymbol{\beta}})/\hat{\sigma}^2$ increases, which divided by u has the F-distribution $F(u,n-u)$ with u and $n-u$ degrees of freedom because of (A22.13)

$$(\boldsymbol{\beta}-\hat{\boldsymbol{\beta}})'\mathbf{X}'\mathbf{X}(\boldsymbol{\beta}-\hat{\boldsymbol{\beta}})/(u\hat{\sigma}^2) \sim F(u,n-u). \tag{312.21}$$

With the same reasoning, which leads to (242.6), the confidence region of content $1-\alpha$ is therefore determined by

$$(\boldsymbol{\beta}-\hat{\boldsymbol{\beta}})'\mathbf{X}'\mathbf{X}(\boldsymbol{\beta}-\hat{\boldsymbol{\beta}})/(u\hat{\sigma}^2) = F_{1-\alpha;u,n-u}, \tag{312.22}$$

where $F_{1-\alpha;u,n-u}$ denotes the upper α-percentage point of the F-distribution with u and n-u degrees of freedom. The confidence region is a hyperellipsoid with the center at $\hat{\boldsymbol{\beta}}$. It is identical with the confidence hyperellipsoid of the standard statistical techniques (Koch 1988a, p.328).

If a confidence region needs to be established for a subset $\boldsymbol{\beta}_{j..k}$ of the parameters $\boldsymbol{\beta}$, we define

$$\boldsymbol{\beta} = (\beta_i),\ \boldsymbol{\beta}_{j..k} = (\beta_l) \text{ and } \hat{\boldsymbol{\beta}} = (\hat{\beta}_i),\ \hat{\boldsymbol{\beta}}_{j..k} = (\hat{\beta}_l)$$
$$\text{for } l\in\{j,j+1,\dots,k\}, \tag{312.23}$$
$$\boldsymbol{\Sigma} = (\sigma_{ij}) = (\mathbf{X}'\mathbf{X})^{-1} \text{ and } \boldsymbol{\Sigma}_{j..k} = (\sigma_{lm}) \text{ for } l,m\in\{j,j+1,\dots,k\}$$

and obtain with (A22.10) the marginal multivariate t-distribution

$$\boldsymbol{\beta}_{j..k}|\mathbf{y} \sim t(\hat{\boldsymbol{\beta}}_{j..k},\hat{\sigma}^2\boldsymbol{\Sigma}_{j..k},n-u). \tag{312.24}$$

This distribution leads to the confidence region of content $1-\alpha$

$$(\boldsymbol{\beta}_{j..k}-\hat{\boldsymbol{\beta}}_{j..k})'\boldsymbol{\Sigma}_{j..k}^{-1}(\boldsymbol{\beta}_{j..k}-\hat{\boldsymbol{\beta}}_{j..k})/((k-j+1)\hat{\sigma}^2) = F_{1-\alpha;k-j+1,n-u}, \tag{312.25}$$

which again is identical with the confidence hyperellipsoid of the standard statistical techniques (Koch 1988a, p.328).

We want to test the point null hypothesis (251.4) generalized by (251.5)

$$H_0 : \mathbf{H}\boldsymbol{\beta} = \mathbf{w} \text{ versus } H_1 : \mathbf{H}\boldsymbol{\beta} \neq \mathbf{w}, \tag{312.26}$$

where the r×u matrix $\mathbf{H}$ has full row rank and is given. The r×1 vector $\mathbf{w}$ is also given. With (312.7) we obtain from (A22.12) the distribution

$$\mathbf{H}\boldsymbol{\beta} \sim t(\mathbf{H}\hat{\boldsymbol{\beta}},\hat{\sigma}^2\mathbf{H}(\mathbf{X}'\mathbf{X})^{-1}\mathbf{H}',n-u) \tag{312.27}$$

and from (A22.13)

$$(\mathbf{H}\boldsymbol{\beta}-\mathbf{H}\hat{\boldsymbol{\beta}})'(\mathbf{H}(\mathbf{X}'\mathbf{X})^{-1}\mathbf{H}')^{-1}(\mathbf{H}\boldsymbol{\beta}-\mathbf{H}\hat{\boldsymbol{\beta}})/(r\hat{\sigma}^2) \sim F(r,n-u). \tag{312.28}$$

We will decide by (252.4), that is by means of the confidence region for $\mathbf{H}\boldsymbol{\beta}$, whether to accept or to reject the null hypothesis H_0 in (312.26). The confidence region for $\mathbf{H}\boldsymbol{\beta}$ follows from (312.28) by

$$(\mathbf{H}\boldsymbol{\beta}-\mathbf{H}\hat{\boldsymbol{\beta}})'(\mathbf{H}(\mathbf{X}'\mathbf{X})^{-1}\mathbf{H}')^{-1}(\mathbf{H}\boldsymbol{\beta}-\mathbf{H}\hat{\boldsymbol{\beta}})/(r\hat{\sigma}^2) = F_{1-\alpha;r,n-u}. \tag{312.29}$$

According to (A22.1) the posterior density for $\mathbf{H}\boldsymbol{\beta}$ increases, when the quadratic form on the left-hand side of (312.29) decreases. Thus, because of (252.4) and $\mathbf{H}\boldsymbol{\beta}=\mathbf{w}$ the null hypothesis H_o in (312.26) is rejected, if

$$(\mathbf{H}\hat{\boldsymbol{\beta}}-\mathbf{w})'(\mathbf{H}(\mathbf{X}'\mathbf{X})^{-1}\mathbf{H}')^{-1}(\mathbf{H}\hat{\boldsymbol{\beta}}-\mathbf{w})/(r\hat{\sigma}^2) > F_{1-\alpha;r,n-u}. \tag{312.30}$$

An equivalent procedure is applied, when the hypothesis (312.26) is tested by the standard statistical test. The quantity on the left-hand side of (312.30) is the test statistic for this test (Koch 1988a, p.308).

Thus, Bayesian inference for the parameter vector $\boldsymbol{\beta}$ of a linear model gives, in the case of noninformative priors, results which are equivalent to the results of the standard statistical techniques.

Example 2: A confidence interval of content $1-\alpha$ shall be determined for the unknown parameter μ of the Example 1 of Section 311, which is continued by Example 1 of this section. We obtain from (312.21)

$$n(\mu-\hat{\mu})^2/\hat{\sigma}^2 \sim F(1,n-1) \tag{312.31}$$

with $\hat{\mu}$ from (312.8)

$$\hat{\mu} = \frac{1}{n}\sum_{i=1}^{n} y_i .$$

From the definition of the upper α-percentage point of the F-distribution the probability follows

$$P(n(\mu-\hat{\mu})^2/\hat{\sigma}^2 < F_{1-\alpha;1,n-1}) = 1-\alpha. \tag{312.32}$$

By a transformation of the variable of the integral from which this probability is computed we find

$$P(\pm\sqrt{n}(\mu-\hat{\mu})/\hat{\sigma} < (F_{1-\alpha;1,n-1})^{1/2}) = 1-\alpha. \tag{312.33}$$

If a random variable x^2 is distributed according to

$$x^2 \sim F(1,n-1), \text{ then } x \sim t(n-1)$$

(Koch 1988a, p.155), where $t(n-1)$ is the t-distribution with $n-1$ degrees of freedom having the density function (A22.6). With the quantity $t_{\alpha;n-1}$ of the t-distribution, which can be taken from a table of the t-distribution and which is equal to the square root of the upper α-percentage point of the F-distribution

$$t_{\alpha;n-1} = (F_{1-\alpha;1,n-1})^{1/2} \tag{312.34}$$

we obtain instead of (312.33)

$$P(-t_{\alpha;n-1} < \sqrt{n}(\mu-\hat{\mu})/\hat{\sigma} < t_{\alpha;n-1}) = 1-\alpha$$

and finally the confidence interval of content $1-\alpha$ for the parameter μ

$$P(\hat{\mu}-(\hat{\sigma}/\sqrt{n})\,t_{\alpha;n-1} < \mu < \hat{\mu} + (\hat{\sigma}/\sqrt{n})\,t_{\alpha;n-1}) = 1-\alpha. \tag{312.35}$$

Since $\hat{\sigma}/\sqrt{n}$ is the standard deviation of $\hat{\mu}$ computed by the estimated variance $\hat{\sigma}^2$ of unit weight, the confidence interval (312.35) is identical with the confidence interval for μ of the standard statistical techniques.

In addition we want to test the hypothesis

$$H_o : \mu = \mu_o \quad \text{versus} \quad H_1 : \mu \neq \mu_o, \tag{312.36}$$

where μ_o is a given constant. We will use (252.4) to decide whether to accept the hypothesis. Thus, if μ_o lies outside the confidence interval (312.35), H_o is rejected. An equivalent condition for μ_o lying outside the confidence interval is obtained from (312.30). With $\mathbf{H}=1$, $\boldsymbol{\beta}=\mu$ from (311.12) and $\mathbf{w}=\mu_o$, the hypothesis (312.36) is represented in the general form (312.26). Thus, we obtain instead of (312.30)

$$\frac{(\hat{\mu}-\mu_o)^2}{\hat{\sigma}^2/n} > F_{1-\alpha;1,n-1}. \tag{312.37}$$

If this inequality is fulfilled, H_o is rejected. The same procedure is applied in the standard statistical techniques to decide whether to accept or to reject the null hypothesis H_o in (312.36). Δ

313 Informative Priors

We will assume that prior information is given for the parameters $\boldsymbol{\beta}$ and σ^2 in the linear model (311.1) by means of the expected value $\boldsymbol{\mu}_p$ and the covariance matrix $\boldsymbol{\Sigma}_\beta$ of the parameter vector $\boldsymbol{\beta}$ and by the expected value σ_p^2 and the variance V_{σ^2} of the parameter σ^2. We introduce the normal-gamma distribution as prior for $\boldsymbol{\beta}$ and $\tau=1/\sigma^2$

$$\boldsymbol{\beta}, \tau \sim NG(\boldsymbol{\mu}, \mathbf{V}, b, p). \tag{313.1}$$

The parameters of this distribution are determined with (224.20) by

$$\boldsymbol{\mu} = \boldsymbol{\mu}_p, \; \mathbf{V} = \boldsymbol{\Sigma}_\beta/\sigma_p^2, \; p = (\sigma_p^2)^2/V_{\sigma^2} + 2, \; b = (p-1)\sigma_p^2. \tag{313.2}$$

The normal-gamma distribution (313.1) is a natural conjugate prior, so that together with the likelihood function (311.9) Bayes' theorem (211.1) gives the posterior distribution for $\boldsymbol{\beta}, \tau$ which is also normal-gamma

$$\boldsymbol{\beta}, \tau|\mathbf{y} \sim NG(\boldsymbol{\mu}_o, \mathbf{V}_o, b_o, p_o). \tag{313.3}$$

The parameters of this distribution are given in (224.10).

The marginal posterior distribution of the parameter vector $\boldsymbol{\beta}$ is therefore the multivariate t-distribution

$$\boldsymbol{\beta}|\mathbf{y} \sim t(\boldsymbol{\mu}_0, b_0\mathbf{V}_0/p_0, 2p_0) \tag{313.4}$$

according to (A23.3). The expected value and the covariance matrix of $\boldsymbol{\beta}$ from (A22.7) give according to (231.5) and (231.7) the Bayes estimate $\hat{\boldsymbol{\beta}}_B$ of $\boldsymbol{\beta}$ and the covariance matrix $\boldsymbol{\Sigma}_{\hat{\beta}}$ of the estimate $\hat{\boldsymbol{\beta}}_B$. Thus, with (224.10)

$$\hat{\boldsymbol{\beta}}_B = \boldsymbol{\mu}_0 = (\mathbf{X}'\mathbf{X}+\mathbf{V}^{-1})^{-1}(\mathbf{X}'\mathbf{y}+\mathbf{V}^{-1}\boldsymbol{\mu}) \tag{313.5}$$

and

$$\boldsymbol{\Sigma}_{\hat{\beta}} = \frac{1}{n+2p-2}\,(2b+(\boldsymbol{\mu}-\boldsymbol{\mu}_0)'\mathbf{V}^{-1}(\boldsymbol{\mu}-\boldsymbol{\mu}_0)+(\mathbf{y}-\mathbf{X}\boldsymbol{\mu}_0)'(\mathbf{y}-\mathbf{X}\boldsymbol{\mu}_0))(\mathbf{X}'\mathbf{X}+\mathbf{V}^{-1})^{-1}. \tag{313.6}$$

The Bayes estimate $\hat{\boldsymbol{\beta}}_B$ and its covariance matrix $\boldsymbol{\Sigma}_{\hat{\beta}}$ shall be also given in the more general model (311.3). By substituting (311.4) we find

$$\hat{\boldsymbol{\beta}}_B = \boldsymbol{\mu}_0 = (\bar{\mathbf{X}}'\mathbf{P}\bar{\mathbf{X}}+\mathbf{V}^{-1})^{-1}(\bar{\mathbf{X}}'\mathbf{P}\bar{\mathbf{y}}+\mathbf{V}^{-1}\boldsymbol{\mu}) \tag{313.7}$$

and

$$\boldsymbol{\Sigma}_{\hat{\beta}} = \frac{1}{n+2p-2}\,(2b+(\boldsymbol{\mu}-\boldsymbol{\mu}_0)'\mathbf{V}^{-1}(\boldsymbol{\mu}-\boldsymbol{\mu}_0) + (\bar{\mathbf{y}}-\bar{\mathbf{X}}\boldsymbol{\mu}_0)'\mathbf{P}(\bar{\mathbf{y}}-\bar{\mathbf{X}}\boldsymbol{\mu}_0))(\bar{\mathbf{X}}'\mathbf{P}\bar{\mathbf{X}}+\mathbf{V}^{-1})^{-1}. \tag{313.8}$$

For easier comparison with the estimates of the mixed model to be presented in Section 333, and for obtaining in special cases computationally more efficient formulas, the Bayes estimate $\hat{\boldsymbol{\beta}}_B$ will be presented in the following form. We apply the two matrix identities (Koch 1988a, p.40)

$$(\bar{\mathbf{X}}'\mathbf{P}\bar{\mathbf{X}}+\mathbf{V}^{-1})^{-1}\bar{\mathbf{X}}'\mathbf{P} = \mathbf{V}\bar{\mathbf{X}}'(\bar{\mathbf{X}}\mathbf{V}\bar{\mathbf{X}}'+\mathbf{P}^{-1})^{-1}$$

$$(\bar{\mathbf{X}}'\mathbf{P}\bar{\mathbf{X}}+\mathbf{V}^{-1})^{-1} = \mathbf{V} - \mathbf{V}\bar{\mathbf{X}}'(\bar{\mathbf{X}}\mathbf{V}\bar{\mathbf{X}}'+\mathbf{P}^{-1})^{-1}\bar{\mathbf{X}}\mathbf{V} \tag{313.9}$$

and obtain

$$\hat{\boldsymbol{\beta}}_B = \boldsymbol{\mu}_0 = \boldsymbol{\mu} + \mathbf{V}\bar{\mathbf{X}}'(\bar{\mathbf{X}}\mathbf{V}\bar{\mathbf{X}}'+\mathbf{P}^{-1})^{-1}(\bar{\mathbf{y}}-\bar{\mathbf{X}}\boldsymbol{\mu}). \tag{313.10}$$

In general $n>u$ holds for the linear model (311.1). However, when using prior information we may have $n<u$. In such a case, (313.10) is computationally more efficient than (313.7), since the inverse in (313.10) is of dimension $n \times n$. If the vector $\boldsymbol{\mu}$ and the matrix $\mathbf{V}$ stemming from the prior information are considered as results of a preceding parameter estimation, (313.10) represents a recursive estimation (Koch 1988a, p.208), which leads for a dynamic system to the Kalman-Bucy filter derived in Section 318.

It is worth mentioning that the Bayes estimate (313.10) is closely related to the estimate of the robust collocation in the presence of weak prior information (Schaffrin 1989)

$$\begin{aligned}\bar{\boldsymbol{\beta}} &= \boldsymbol{\mu}\hat{\omega} + \mathbf{V}\bar{\mathbf{X}}'(\bar{\mathbf{X}}\mathbf{V}\bar{\mathbf{X}}'+\mathbf{P}^{-1})^{-1}(\bar{\mathbf{y}}-\bar{\mathbf{X}}\boldsymbol{\mu}\hat{\omega}) \\ &= \boldsymbol{\mu}\hat{\omega} + \mathbf{V}(\mathbf{I}+\bar{\mathbf{X}}'\mathbf{P}\bar{\mathbf{X}}\mathbf{V})^{-1}\bar{\mathbf{X}}'\mathbf{P}(\bar{\mathbf{y}}-\bar{\mathbf{X}}\boldsymbol{\mu}\hat{\omega})\end{aligned} \qquad (313.11)$$

with

$$\hat{\omega} = \frac{\boldsymbol{\mu}'\bar{\mathbf{X}}'(\bar{\mathbf{X}}\mathbf{V}\bar{\mathbf{X}}'+\mathbf{P}^{-1})^{-1}\bar{\mathbf{y}}}{\boldsymbol{\mu}'\bar{\mathbf{X}}'(\bar{\mathbf{X}}\mathbf{V}\bar{\mathbf{X}}'+\mathbf{P}^{-1})^{-1}\bar{\mathbf{X}}\boldsymbol{\mu}} = \frac{\boldsymbol{\mu}'(\mathbf{I}+\bar{\mathbf{X}}'\mathbf{P}\bar{\mathbf{X}}\mathbf{V})^{-1}\bar{\mathbf{X}}'\mathbf{P}\bar{\mathbf{y}}}{\boldsymbol{\mu}'(\mathbf{I}+\bar{\mathbf{X}}'\mathbf{P}\bar{\mathbf{X}}\mathbf{V})^{-1}\bar{\mathbf{X}}'\mathbf{P}\bar{\mathbf{X}}\boldsymbol{\mu}},$$

where $\bar{\boldsymbol{\beta}}$ denotes the estimate of $\boldsymbol{\beta}$ and where the remaining quantities have the same meaning as in (313.10). The prior information $\boldsymbol{\mu}$ in (313.11) is considered as being weak, since it needs to be scaled by a factor ω, which is estimated by $\hat{\omega}$. The appropriate model for deriving $\bar{\boldsymbol{\beta}}$ and $\hat{\omega}$ in (313.11) is a random effects model of the standard statistical techniques, but $\bar{\boldsymbol{\beta}}$ and $\hat{\omega}$ may also be obtained by the following Gauss-Markoff model of the standard statistical techniques

$$\begin{bmatrix}\bar{\mathbf{y}}\\ \mathbf{0}\end{bmatrix} + \begin{bmatrix}\mathbf{e}\\ \mathbf{v}\end{bmatrix} = \begin{bmatrix}\bar{\mathbf{X}} & \mathbf{0}\\ \mathbf{I} & -\boldsymbol{\mu}\end{bmatrix}\begin{bmatrix}\boldsymbol{\beta}\\ \omega\end{bmatrix} \quad \text{with} \quad D(\begin{bmatrix}\bar{\mathbf{y}}\\ \mathbf{0}\end{bmatrix}) = \sigma^2\begin{bmatrix}\mathbf{P}^{-1} & \mathbf{0}\\ \mathbf{0} & \mathbf{V}\end{bmatrix}. \qquad (313.12)$$

This can be shown by solving the matrix of normal equations resulting from (313.12) for $\hat{\omega}$ by means of the inverse of a block matrix (Koch 1988a, p.39). By substituting $\hat{\omega}$ the normal equations are then solved for $\bar{\boldsymbol{\beta}}$. For obtaining (313.11) the two matrix identities (313.9) are applied.

To compare the result (313.5) with the result of the standard statistical techniques, we assume that in addition to the observation equations $\mathbf{y}+\mathbf{e}=\mathbf{X}\boldsymbol{\beta}$ of the Gauss-Markoff model, the unknown parameters $\boldsymbol{\beta}$ have been independently observed by the vector $\boldsymbol{\mu}$ with the error vector $\mathbf{v}$ and the weight matrix $\mathbf{V}^{-1}$. Thus,

$$\begin{bmatrix}\mathbf{y}\\ \boldsymbol{\mu}\end{bmatrix} + \begin{bmatrix}\mathbf{e}\\ \mathbf{v}\end{bmatrix} = \begin{bmatrix}\mathbf{X}\\ \mathbf{I}\end{bmatrix}\boldsymbol{\beta} \quad \text{with} \quad D(\begin{bmatrix}\mathbf{y}\\ \boldsymbol{\mu}\end{bmatrix}) = \sigma^2\begin{bmatrix}\mathbf{I} & \mathbf{0}\\ \mathbf{0} & \mathbf{V}\end{bmatrix}, \qquad (313.13)$$

which gives with (224.4) the estimate

$$\hat{\boldsymbol{\beta}} = \boldsymbol{\mu}_0 = (\mathbf{X}'\mathbf{X}+\mathbf{V}^{-1})^{-1}(\mathbf{X}'\mathbf{y}+\mathbf{V}^{-1}\boldsymbol{\mu}). \qquad (313.14)$$

It is identical with (313.5). If the prior information is interpreted as data, the Bayes estimate $\hat{\boldsymbol{\beta}}_B$ of the parameter vector $\boldsymbol{\beta}$ and the estimate of the standard statistical techniques are identical. The same, of course, holds true for (313.7).

With (313.3) the marginal posterior distribution of the weight parameter τ is according to (A23.4) the gamma distribution

$$\tau|\mathbf{y} \sim G(b_o, p_o). \tag{313.15}$$

The variance σ^2 of unit weight therefore has the inverted gamma distribution (A13.1) with the expected value and variance from (A13.2), leading to the Bayes estimate $\hat{\sigma}^2_B$ of σ^2 and to the variance $V(\hat{\sigma}^2_B)$ of $\hat{\sigma}^2_B$. Thus, with (224.10)

$$\hat{\sigma}^2_B = \frac{1}{n+2p-2}\,(2b+(\boldsymbol{\mu}-\boldsymbol{\mu}_o)'\mathbf{V}^{-1}(\boldsymbol{\mu}-\boldsymbol{\mu}_o)+(\mathbf{y}-\mathbf{X}\boldsymbol{\mu}_o)'(\mathbf{y}-\mathbf{X}\boldsymbol{\mu}_o)) \tag{313.16}$$

and

$$V(\hat{\sigma}^2_B) = \frac{2(\hat{\sigma}^2_B)^2}{n+2p-4}\,. \tag{313.17}$$

To compare the Bayes estimate $\hat{\sigma}^2_B$ of σ^2 with the estimate $\hat{\sigma}^2$ of the standard statistical techniques, we obtain with (224.4), where in the expression for $\hat{\sigma}^2$ the weight matrix from (313.13) has to be introduced (Koch 1988a, p.192),

$$\hat{\sigma}^2 = \frac{1}{n}\,[(\mathbf{y}-\mathbf{X}\boldsymbol{\mu}_o)',\ (\boldsymbol{\mu}-\boldsymbol{\mu}_o)']\begin{bmatrix}\mathbf{I} & \mathbf{0}\\ \mathbf{0} & \mathbf{V}^{-1}\end{bmatrix}\begin{bmatrix}\mathbf{y}-\mathbf{X}\boldsymbol{\mu}_o\\ \boldsymbol{\mu}-\boldsymbol{\mu}_o\end{bmatrix}$$

$$= \frac{1}{n}\,((\boldsymbol{\mu}-\boldsymbol{\mu}_o)'\mathbf{V}^{-1}(\boldsymbol{\mu}-\boldsymbol{\mu}_o)+(\mathbf{y}-\mathbf{X}\boldsymbol{\mu}_o)'(\mathbf{y}-\mathbf{X}\boldsymbol{\mu}_o)). \tag{313.18}$$

By substituting b and p from (313.2) in (313.16), we recognize that the prior information on σ^2 enters the estimate in comparison to (313.18). In addition the number of degrees of freedom differs.

By means of the substitution

$$\mathbf{V}^{-1} \to \mathbf{0},\quad b \to 0,\quad p \to -u/2 \tag{313.19}$$

the parameters of the normal-gamma distribution (313.3) for $\boldsymbol{\beta}$ and τ result with (224.4) and (224.10) by

$$\boldsymbol{\mu}_o = \hat{\boldsymbol{\beta}} = (\mathbf{X}'\mathbf{X})^{-1}\mathbf{X}'\mathbf{y}$$
$$\mathbf{V}_o = (\mathbf{X}'\mathbf{X})^{-1}$$
$$b_o = (n-u)\hat{\sigma}^2/2$$
$$p_o = (n-u)/2.$$

The distribution for $\boldsymbol{\beta}$ and τ follows with

$$\boldsymbol{\beta},\tau|\mathbf{y} \sim NG(\hat{\boldsymbol{\beta}}, (\mathbf{X}'\mathbf{X})^{-1}, (n-u)\hat{\sigma}^2/2, (n-u)/2),$$

which is the normal-gamma distribution (312.6) of $\boldsymbol{\beta}$ and τ in the case of noninformative priors. Hence, (313.19) transforms the results with informative priors to the results with noninformative priors. One has to be aware, however, that substituting (313.19) in the informative prior density (313.1) for $\boldsymbol{\beta}$ and τ does not lead to the noninformative priors (312.2) and (312.4) for $\boldsymbol{\beta}$ and τ.

Example 1: Again the unknown parameters μ and σ^2 of the linear model (311.11) of Example 1 of Section 311 shall be estimated. But instead of introducing noninformative priors as in Example 1 of Section 312, an informative prior is assumed for μ and σ^2. It is given with $\tau=1/\sigma^2$ by the normal-gamma distribution (313.1)

$$\mu,\tau \sim NG(\bar{\mu},V,b,p) . \tag{313.20}$$

The parameters of this distribution are determined by the prior information on the expected values and the variances of μ and σ^2

$$E(\mu) = \mu_p ,\ V(\mu) = \sigma_\mu^2$$
$$E(\sigma^2) = \sigma_p^2 ,\ V(\sigma^2) = V_{\sigma^2} ,$$

which gives with (313.2)

$$\bar{\mu} = \mu_p ,\ V = \sigma_\mu^2/\sigma_p^2 ,\ p = (\sigma_p^2)^2/V_{\sigma^2} + 2 ,\ b = (p-1)\sigma_p^2 . \tag{313.21}$$

The likelihood function of this example is given by (311.14). It is identical with the likelihood function (224.21) of the Example 1 of Section 224. In addition the prior distribution (313.20) for μ and σ^2 with the parameters from (313.21) is identical with the prior distribution (224.22) with the parameters (224.24). Hence, the posterior distribution for μ and σ^2 of this example is given by the posterior distribution (224.25). The Bayes estimates for μ and σ^2 obtained from (313.5) and (313.16) and their variances from (313.6) and (313.17) therefore follow with (224.27) to (224.30) and with (224.32) to (224.35). **Δ**

314 Prediction of Data

If we want to predict m observations of a linear model, we may either start from the formulation (311.1) $E(\mathbf{y}|\boldsymbol{\beta})=\mathbf{X}\boldsymbol{\beta}$ of the linear model or from the equivalent formulation (311.2) $\mathbf{y}=\mathbf{X}\boldsymbol{\beta}-\mathbf{e}$. In the first case we predict with

$$\mathbf{y}_p = E(\mathbf{y}^*|\boldsymbol{\beta}) = \mathbf{X}^*\boldsymbol{\beta} \tag{314.1}$$

the m×1 vector $\mathbf{y}_p$ of observations as the expected value of the m×1 vector $\mathbf{y}^*$ given $\boldsymbol{\beta}$. The matrix $\mathbf{X}^*$ denotes the m×u matrix, which has to be known for the prediction.

The distribution for the unobserved data vector $\mathbf{y}_p$ can be readily given. If we assume as the posterior distribution for the parameter vector $\boldsymbol{\beta}$ the multivariate t-distribution (313.4), which results from the informative prior (313.1) for the parameters $\boldsymbol{\beta}$ and τ, the distribution of the linear function $\mathbf{y}_p=\mathbf{X}^*\boldsymbol{\beta}$ of $\boldsymbol{\beta}$ follows with (A22.12) by

$$\mathbf{y}_p|\mathbf{y} \sim t(\mathbf{X}^*\boldsymbol{\mu}_o, b_o\mathbf{X}^*\mathbf{V}_o\mathbf{X}^{*\prime}/p_o, 2p_o). \tag{314.2}$$

All inferential tasks for the predicted observation vector $\mathbf{y}_p$ can be solved by this distribution. For instance the Bayes estimate $\hat{\mathbf{y}}_{pB}$ of $\mathbf{y}_p$ and the covariance matrix $\boldsymbol{\Sigma}_{\hat{y}p}$ of the estimate $\hat{\mathbf{y}}_{pB}$ are obtained from (231.5), (231.7), (313.5), (313.6) and (A22.7) by

$$\hat{\mathbf{y}}_{pB} = \mathbf{X}^*\boldsymbol{\mu}_o = \mathbf{X}^*(\mathbf{X}'\mathbf{X}+\mathbf{V}^{-1})^{-1}(\mathbf{X}'\mathbf{y}+\mathbf{V}^{-1}\boldsymbol{\mu}) \tag{314.3}$$

and

$$\boldsymbol{\Sigma}_{\hat{y}p} = \frac{1}{n+2p-2}[2b+(\boldsymbol{\mu}-\boldsymbol{\mu}_o)'\mathbf{V}^{-1}(\boldsymbol{\mu}-\boldsymbol{\mu}_o)+(\mathbf{y}-\mathbf{X}\boldsymbol{\mu}_o)'(\mathbf{y}-\mathbf{X}\boldsymbol{\mu}_o)]\mathbf{X}^*(\mathbf{X}'\mathbf{X}+\mathbf{V}^{-1})^{-1}\mathbf{X}^{*\prime}. \tag{314.4}$$

By substituting (313.19) in the density function defined by (314.2) the results are obtained which are based on noninformative priors for the parameters $\boldsymbol{\beta}$ and τ.

If we do not want to predict observations as expected values but as actual observations, we start as mentioned above from the formulation (311.2) $\mathbf{y}=\mathbf{X}\boldsymbol{\beta}-\mathbf{e}$ of the linear model and predict the m×1 vector $\mathbf{y}_u$ of unobserved data by

$$\mathbf{y}_u = \mathbf{X}^*\boldsymbol{\beta} - \mathbf{u}, \tag{314.5}$$

where the m×u matrix $\mathbf{X}^*$ has to be known and where $\mathbf{u}$ denotes an m×1 vector of unknown errors.

The predictive density function of the unobserved data vector $\mathbf{y}_u$ follows from (262.2) by

$$p(\mathbf{y}_u|\mathbf{y}) = \int_0^\infty \int_{-\infty}^\infty \dots \int_{-\infty}^\infty p(\mathbf{y}_u|\boldsymbol{\beta},\tau,\mathbf{y})\ p(\boldsymbol{\beta},\tau|\mathbf{y})\ d\beta_1\dots d\beta_u d\tau, \tag{314.6}$$

since the parameters of the linear model (311.1) are $\boldsymbol{\beta}$ and τ. In agreement with the normal distribution (311.8) for the observation vector $\mathbf{y}$ we assume the normal distribution for the predicted vector $\mathbf{y}_u$ under the condition that $\boldsymbol{\beta}$, τ and $\mathbf{y}$ are given

$$\mathbf{y}_u|\boldsymbol{\beta},\tau,\mathbf{y} \sim N(\mathbf{X}^*\boldsymbol{\beta}, \tau^{-1}\mathbf{I}). \tag{314.7}$$

This leads to the density

$$p(\mathbf{y}_u|\boldsymbol{\beta},\tau,\mathbf{y}) \propto \tau^{m/2}\exp[-\frac{\tau}{2}(\mathbf{y}_u-\mathbf{X}^*\boldsymbol{\beta})'(\mathbf{y}_u-\mathbf{X}^*\boldsymbol{\beta})]. \tag{314.8}$$

The posterior density $p(\boldsymbol{\beta},\tau|\mathbf{y})$ in (314.6) shall be given by the normal-gamma distribution (313.3), which is based on the normal-gamma distribution (313.1) as informative prior distribution and the likelihood function (311.9). Thus, with (A23.1)

$$p(\boldsymbol{\beta},\tau|\mathbf{y}) \propto \tau^{u/2+p-1}\exp\{-\frac{\tau}{2}[2b+(\boldsymbol{\beta}-\boldsymbol{\mu})'\mathbf{V}^{-1}(\boldsymbol{\beta}-\boldsymbol{\mu})]\}$$
$$\tau^{n/2}\exp[-\frac{\tau}{2}(\mathbf{y}-\mathbf{X}\boldsymbol{\beta})'(\mathbf{y}-\mathbf{X}\boldsymbol{\beta})]. \qquad (314.9)$$

By substituting (314.8) and (314.9) in (314.6) we obtain

$$p(\mathbf{y}_u|\mathbf{y}) = \int_0^\infty p(\mathbf{y}_u,\tau|\mathbf{y})d\tau \qquad (314.10)$$

with

$$p(\mathbf{y}_u,\tau|\mathbf{y}) \propto \int_{-\infty}^{\infty}\ldots\int_{-\infty}^{\infty} \tau^{(m+n+u)/2+p-1}\exp\{-\frac{\tau}{2}[2b+(\mathbf{y}_u-\mathbf{X}^*\boldsymbol{\beta})'(\mathbf{y}_u-\mathbf{X}^*\boldsymbol{\beta})$$
$$+(\boldsymbol{\beta}-\boldsymbol{\mu})'\mathbf{V}^{-1}(\boldsymbol{\beta}-\boldsymbol{\mu}) + (\mathbf{y}-\mathbf{X}\boldsymbol{\beta})'(\mathbf{y}-\mathbf{X}\boldsymbol{\beta})]\}d\beta_1\ldots d\beta_u, \qquad (314.11)$$

where $p(\mathbf{y}_u,\tau|\mathbf{y})$ denotes the joint density of $\mathbf{y}_u$ and τ given $\mathbf{y}$. We complete the squares on $\boldsymbol{\beta}$ in the exponent of (314.11) and find

$$2b + \mathbf{y}_u'\mathbf{y}_u + \boldsymbol{\mu}'\mathbf{V}^{-1}\boldsymbol{\mu} + \mathbf{y}'\mathbf{y} + \boldsymbol{\beta}'\mathbf{M}\boldsymbol{\beta} - 2\boldsymbol{\beta}'(\mathbf{X}'\mathbf{y}+\mathbf{X}^{*\prime}\mathbf{y}_u+\mathbf{V}^{-1}\boldsymbol{\mu})$$
$$= 2b + \mathbf{y}_u'\mathbf{y}_u + \boldsymbol{\mu}'\mathbf{V}^{-1}\boldsymbol{\mu} + \mathbf{y}'\mathbf{y} - (\mathbf{X}'\mathbf{y}+\mathbf{X}^{*\prime}\mathbf{y}_u+\mathbf{V}^{-1}\boldsymbol{\mu})'\mathbf{M}^{-1}(\mathbf{X}'\mathbf{y}+\mathbf{X}^{*\prime}\mathbf{y}_u+\mathbf{V}^{-1}\boldsymbol{\mu})$$
$$+ (\boldsymbol{\beta}-\mathbf{M}^{-1}(\mathbf{X}'\mathbf{y}+\mathbf{X}^{*\prime}\mathbf{y}_u+\mathbf{V}^{-1}\boldsymbol{\mu}))'\mathbf{M}(\boldsymbol{\beta}-\mathbf{M}^{-1}(\mathbf{X}'\mathbf{y}+\mathbf{X}^{*\prime}\mathbf{y}_u+\mathbf{V}^{-1}\boldsymbol{\mu}))$$

with

$$\mathbf{M} = \mathbf{X}'\mathbf{X} + \mathbf{X}^{*\prime}\mathbf{X}^* + \mathbf{V}^{-1}. \qquad (314.12)$$

The integral in (314.11) with respect to $\boldsymbol{\beta}$ can now be solved and yields, because of (A21.2),

$$\int_{-\infty}^{\infty}\ldots\int_{-\infty}^{\infty} \exp\{-\frac{\tau}{2}[\boldsymbol{\beta}-\mathbf{M}^{-1}(\mathbf{X}'\mathbf{y}+\mathbf{X}^{*\prime}\mathbf{y}_u+\mathbf{V}^{-1}\boldsymbol{\mu}))'\mathbf{M}(\boldsymbol{\beta}-\mathbf{M}^{-1}(\mathbf{X}'\mathbf{y}+\mathbf{X}^{*\prime}\mathbf{y}_u+\mathbf{V}^{-1}\boldsymbol{\mu})]\}$$
$$d\beta_1\ldots d\beta_u = (2\pi)^{u/2}(\det(\frac{1}{\tau}\mathbf{M}^{-1}))^{1/2} \propto \tau^{-u/2}.$$

The joint density $p(\mathbf{y}_u,\tau|\mathbf{y})$ of $\mathbf{y}_u$ and τ follows with

$$p(\mathbf{y}_u,\tau|\mathbf{y}) \propto \tau^{(m+n)/2+p-1}\exp\{-\frac{\tau}{2}[2b+\mathbf{y}_u'\mathbf{y}_u+\boldsymbol{\mu}'\mathbf{V}^{-1}\boldsymbol{\mu}+\mathbf{y}'\mathbf{y}$$
$$-(\mathbf{X}'\mathbf{y}+\mathbf{X}^{*\prime}\mathbf{y}_u+\mathbf{V}^{-1}\boldsymbol{\mu})'\mathbf{M}^{-1}(\mathbf{X}'\mathbf{y}+\mathbf{X}^{*\prime}\mathbf{y}_u+\mathbf{V}^{-1}\boldsymbol{\mu})]\}. \qquad (314.13)$$

The exponent of this density shall be denoted by E. We complete the squares on $\mathbf{y}_u$ in E and find

$$E = 2b + \mathbf{y}'\mathbf{y} - \mathbf{y}'\mathbf{X}\mathbf{M}^{-1}\mathbf{X}'\mathbf{y} + \boldsymbol{\mu}'\mathbf{V}^{-1}\boldsymbol{\mu} - \boldsymbol{\mu}'\mathbf{V}^{-1}\mathbf{M}^{-1}\mathbf{V}^{-1}\boldsymbol{\mu}$$
$$+ \mathbf{y}_u'(\mathbf{I}-\mathbf{X}^*\mathbf{M}^{-1}\mathbf{X}^{*\prime})\mathbf{y}_u - 2\mathbf{y}_u'\mathbf{X}^*\mathbf{M}^{-1}(\mathbf{X}'\mathbf{y}+\mathbf{V}^{-1}\boldsymbol{\mu}) - 2\mathbf{y}'\mathbf{X}\mathbf{M}^{-1}\mathbf{V}^{-1}\boldsymbol{\mu}$$

$$= 2b + \mathbf{y}'\mathbf{y} + \boldsymbol{\mu}'\mathbf{V}^{-1}\boldsymbol{\mu} - (\mathbf{X}'\mathbf{y}+\mathbf{V}^{-1}\boldsymbol{\mu})'\mathbf{M}^{-1}(\mathbf{X}'\mathbf{y}+\mathbf{V}^{-1}\boldsymbol{\mu})$$
$$- (\mathbf{X}^*\mathbf{M}^{-1}(\mathbf{X}'\mathbf{y}+\mathbf{V}^{-1}\boldsymbol{\mu}))'(\mathbf{I}-\mathbf{X}^*\mathbf{M}^{-1}\mathbf{X}^{*\prime})^{-1}(\mathbf{X}^*\mathbf{M}^{-1}(\mathbf{X}'\mathbf{y}+\mathbf{V}^{-1}\boldsymbol{\mu}))$$
$$+ (\mathbf{y}_u-(\mathbf{I}-\mathbf{X}^*\mathbf{M}^{-1}\mathbf{X}^{*\prime})^{-1}\mathbf{X}^*\mathbf{M}^{-1}(\mathbf{X}'\mathbf{y}+\mathbf{V}^{-1}\boldsymbol{\mu}))'(\mathbf{I}-\mathbf{X}^*\mathbf{M}^{-1}\mathbf{X}^{*\prime})$$
$$(\mathbf{y}_u-(\mathbf{I}-\mathbf{X}^*\mathbf{M}^{-1}\mathbf{X}^{*\prime})^{-1}\mathbf{X}^*\mathbf{M}^{-1}(\mathbf{X}'\mathbf{y}+\mathbf{V}^{-1}\boldsymbol{\mu})). \tag{314.14}$$

By substituting (314.14) in (314.13) we recognize with (A23.1) the joint density function $p(\mathbf{y}_u,\tau|\mathbf{y})$ as a normal-gamma density. The parameters of this distribution can be simplified first by means of the matrix identity (Koch 1988a, p.40) and then by (224.10) and (314.12)

$$(\mathbf{I}-\mathbf{X}^*\mathbf{M}^{-1}\mathbf{X}^{*\prime})^{-1}\mathbf{X}^*\mathbf{M}^{-1}(\mathbf{X}'\mathbf{y}+\mathbf{V}^{-1}\boldsymbol{\mu}) = \mathbf{X}^*(\mathbf{M}-\mathbf{X}^{*\prime}\mathbf{X}^*)^{-1}(\mathbf{X}'\mathbf{y}+\mathbf{V}^{-1}\boldsymbol{\mu})$$
$$= \mathbf{X}^*(\mathbf{X}'\mathbf{X}+\mathbf{V}^{-1})^{-1}(\mathbf{X}'\mathbf{y}+\mathbf{V}^{-1}\boldsymbol{\mu})$$
$$= \mathbf{X}^*\boldsymbol{\mu}_o. \tag{314.15}$$

By applying the same matrix identity we find in addition

$$-(\mathbf{X}'\mathbf{y}+\mathbf{V}^{-1}\boldsymbol{\mu})'\mathbf{M}^{-1}(\mathbf{X}'\mathbf{y}+\mathbf{V}^{-1}\boldsymbol{\mu}) - (\mathbf{X}^*\mathbf{M}^{-1}(\mathbf{X}'\mathbf{y}+\mathbf{V}^{-1}\boldsymbol{\mu}))'(\mathbf{I}-\mathbf{X}^*\mathbf{M}^{-1}\mathbf{X}^{*\prime})^{-1}$$
$$\mathbf{X}^*\mathbf{M}^{-1}(\mathbf{X}'\mathbf{y}+\mathbf{V}^{-1}\boldsymbol{\mu})$$
$$= -(\mathbf{X}'\mathbf{y}+\mathbf{V}^{-1}\boldsymbol{\mu})'\mathbf{M}^{-1}(\mathbf{X}'\mathbf{y}+\mathbf{V}^{-1}\boldsymbol{\mu}) - (\mathbf{X}'\mathbf{y}+\mathbf{V}^{-1}\boldsymbol{\mu})'\mathbf{M}^{-1}\mathbf{X}^{*\prime}\mathbf{X}^*$$
$$(\mathbf{X}'\mathbf{X}+\mathbf{V}^{-1})^{-1}(\mathbf{X}'\mathbf{y}+\mathbf{V}^{-1}\boldsymbol{\mu})$$
$$= -(\mathbf{X}'\mathbf{y}+\mathbf{V}^{-1}\boldsymbol{\mu})'[\mathbf{M}^{-1}+\mathbf{M}^{-1}\mathbf{X}^{*\prime}\mathbf{X}^*(\mathbf{X}'\mathbf{X}+\mathbf{V}^{-1})^{-1}](\mathbf{X}'\mathbf{y}+\mathbf{V}^{-1}\boldsymbol{\mu})$$
$$= -(\mathbf{X}'\mathbf{y}+\mathbf{V}^{-1}\boldsymbol{\mu})'\mathbf{M}^{-1}[\mathbf{X}'\mathbf{X}+\mathbf{V}^{-1}+\mathbf{X}^{*\prime}\mathbf{X}^*](\mathbf{X}'\mathbf{X}+\mathbf{V}^{-1})^{-1}(\mathbf{X}'\mathbf{y}+\mathbf{V}^{-1}\boldsymbol{\mu})$$
$$= -(\mathbf{X}'\mathbf{y}+\mathbf{V}^{-1}\boldsymbol{\mu})'(\mathbf{X}'\mathbf{X}+\mathbf{V}^{-1})^{-1}(\mathbf{X}'\mathbf{y}+\mathbf{V}^{-1}\boldsymbol{\mu})$$
$$= -\boldsymbol{\mu}_o'(\mathbf{X}'\mathbf{X}+\mathbf{V}^{-1})\boldsymbol{\mu}_o. \tag{314.16}$$

We substitute (314.15) and (314.16) in (314.14) and obtain

$$E = 2b + \mathbf{y}'\mathbf{y} + \boldsymbol{\mu}'\mathbf{V}^{-1}\boldsymbol{\mu} - \boldsymbol{\mu}_o'(\mathbf{X}'\mathbf{X}+\mathbf{V}^{-1})\boldsymbol{\mu}_o$$
$$+ (\mathbf{y}_u-\mathbf{X}^*\boldsymbol{\mu}_o)'(\mathbf{I}-\mathbf{X}^*\mathbf{M}^{-1}\mathbf{X}^{*\prime})(\mathbf{y}_u-\mathbf{X}^*\boldsymbol{\mu}_o). \tag{314.17}$$

Finally we substitute (224.11) in (314.17) and find instead of (314.13)

$$p(\mathbf{y}_u,\tau|\mathbf{y}) \propto \tau^{(m+n)/2+p-1}\exp\{-\frac{\tau}{2}[2b+(\boldsymbol{\mu}-\boldsymbol{\mu}_o)'\mathbf{V}^{-1}(\boldsymbol{\mu}-\boldsymbol{\mu}_o)$$
$$+ (\mathbf{y}-\mathbf{X}\boldsymbol{\mu}_o)'(\mathbf{y}-\mathbf{X}\boldsymbol{\mu}_o) + (\mathbf{y}_u-\mathbf{X}^*\boldsymbol{\mu}_o)'(\mathbf{I}-\mathbf{X}^*\mathbf{M}^{-1}\mathbf{X}^{*\prime})(\mathbf{y}_u-\mathbf{X}^*\boldsymbol{\mu}_o)]\}. \tag{314.18}$$

By comparing this density with (A23.1) and by applying the matrix identity (Koch 1988a, p.40) and (224.10)

$$(\mathbf{I}-\mathbf{X}^*\mathbf{M}^{-1}\mathbf{X}^{*\prime})^{-1} = \mathbf{I} + \mathbf{X}^*(\mathbf{M}-\mathbf{X}^{*\prime}\mathbf{X}^*)^{-1}\mathbf{X}^{*\prime}$$
$$= \mathbf{I} + \mathbf{X}^*\mathbf{V}_o\mathbf{X}^{*\prime} \quad (314.19)$$

we find the joint distribution for $\mathbf{y}_u$ and τ given $\mathbf{y}$

$$\mathbf{y}_u, \tau|\mathbf{y} \sim NG(\mathbf{X}^*\boldsymbol{\mu}_o, \mathbf{I}+\mathbf{X}^*\mathbf{V}_o\mathbf{X}^{*\prime}, b_o, p_o). \quad (314.20)$$

The marginal distribution for $\mathbf{y}_u$, which gives according to (314.6) the predictive distribution for the unobserved data vector $\mathbf{y}_u$, follows from (A23.3) by

$$\mathbf{y}_u|\mathbf{y} \sim t(\mathbf{X}^*\boldsymbol{\mu}_o, b_o(\mathbf{I}+\mathbf{X}^*\mathbf{V}_o\mathbf{X}^{*\prime})/p_o, 2p_o). \quad (314.21)$$

All inferential problems arising for the predicted observation vector $\mathbf{y}_u$ can be solved by this distribution. The Bayes estimate $\hat{\mathbf{y}}_{uB}$ of $\mathbf{y}_u$, for instance, and its covariance matrix $\boldsymbol{\Sigma}_{\hat{y}u}$ follow from (231.5), (231.7) and (A22.7) by

$$\hat{\mathbf{y}}_{uB} = \mathbf{X}^*\boldsymbol{\mu}_o = \mathbf{X}^*(\mathbf{X}'\mathbf{X}+\mathbf{V}^{-1})^{-1}(\mathbf{X}'\mathbf{y}+\mathbf{V}^{-1}\boldsymbol{\mu}) \quad (314.22)$$

and

$$\boldsymbol{\Sigma}_{\hat{y}u} = \frac{1}{n+2p-2}\,[2b+(\boldsymbol{\mu}-\boldsymbol{\mu}_o)'\mathbf{V}^{-1}(\boldsymbol{\mu}-\boldsymbol{\mu}_o)+(\mathbf{y}-\mathbf{X}\boldsymbol{\mu}_o)'(\mathbf{y}-\mathbf{X}\boldsymbol{\mu}_o)]$$
$$(\mathbf{I}+\mathbf{X}^*(\mathbf{X}'\mathbf{X}+\mathbf{V}^{-1})^{-1}\mathbf{X}^{*\prime}). \quad (314.23)$$

By substituting (313.19) in the density function defined by (314.21) the predictive density function is obtained for the case that noninformative priors are chosen for the parameters $\boldsymbol{\beta}$ and τ.

If we compare the distribution (314.2) for the unobserved data vector $\mathbf{y}_p$ predicted as the expected value of the observations with the predictive distribution (314.21) of the unobserved data vector $\mathbf{y}_u$ predicted as an actual observation vector, we see that the distributions agree except for the parameter which governs the covariance matrix of the Bayes estimate. The covariance matrix (314.4) gives smaller variances than (314.23). This is understandable, since to the linear combination $\mathbf{X}^*\boldsymbol{\beta}$ in (314.1), which predicts $\mathbf{y}_p$, the error vector $\mathbf{u}$ enters (314.5) to predict $\mathbf{y}_u$.

315 Linear Models Not of Full Rank

It has been assumed so far that the matrix $\mathbf{X}$ of coefficients of the linear model has full column rank. We will now allow a rank deficiency for $\mathbf{X}$ and define correspondingly to (311.1) the *linear model not of full rank*

$$E(\mathbf{y}|\boldsymbol{\beta}) = \mathbf{X}\boldsymbol{\beta} \ \text{with} \ \operatorname{rank}\mathbf{X} = q < u \ \text{and} \ D(\mathbf{y}|\sigma^2) = \sigma^2\mathbf{I}. \quad (315.1)$$

Again let the vector $\mathbf{y}$ of observations be normally distributed according to (311.5), so that the likelihood functions (311.6) and (311.9) are valid.

Since with $\mathrm{rank}\mathbf{X}=\mathrm{rank}\mathbf{X}'\mathbf{X}=q$ (Koch 1988a, p.42) the matrix $\mathbf{X}'\mathbf{X}$ of normal equations is singular, the estimates (224.4) of the unknown parameters $\boldsymbol{\beta}$ and σ^2 cannot be computed. We therefore replace the parameter vector $\boldsymbol{\beta}$ by the projected parameter vector $\boldsymbol{\beta}_b$, which in contrast to $\boldsymbol{\beta}$ represents an estimable function (Koch 1988a, p.217)

$$\boldsymbol{\beta}_b = (\mathbf{X}'\mathbf{X})^-_{rs}\mathbf{X}'\mathbf{X}\boldsymbol{\beta}. \tag{315.2}$$

The matrix $(\mathbf{X}'\mathbf{X})^-_{rs}$ is a symmetrical reflexive generalized inverse of $\mathbf{X}'\mathbf{X}$. Because of (Koch 1988a, p.60)

$$\mathbf{X}\boldsymbol{\beta}_b = \mathbf{X}(\mathbf{X}'\mathbf{X})^-_{rs}\mathbf{X}'\mathbf{X}\boldsymbol{\beta} = \mathbf{X}\boldsymbol{\beta} \tag{315.3}$$

the model (315.1) can be rewritten in terms of $\boldsymbol{\beta}_b$. The estimates of $\boldsymbol{\beta}_b$ and σ^2 follow with (Koch 1988a, p.214, 217)

$$\hat{\boldsymbol{\beta}}_b = (\mathbf{X}'\mathbf{X})^-_{rs}\mathbf{X}'\mathbf{y} \quad \text{and} \quad \hat{\sigma}^2 = \frac{1}{n-q}(\mathbf{y}-\mathbf{X}\hat{\boldsymbol{\beta}}_b)'(\mathbf{y}-\mathbf{X}\hat{\boldsymbol{\beta}}_b). \tag{315.4}$$

On completing the square of the exponent of the likelihood function (311.9) with $\boldsymbol{\beta}_b$ as parameter we obtain with (315.3) and (315.4) correspondingly to (224.5)

$$(\mathbf{y}-\mathbf{X}\boldsymbol{\beta}_b)'(\mathbf{y}-\mathbf{X}\boldsymbol{\beta}_b) = (\mathbf{y}-\mathbf{X}\hat{\boldsymbol{\beta}}_b)'(\mathbf{y}-\mathbf{X}\hat{\boldsymbol{\beta}}_b) + (\boldsymbol{\beta}_b-\hat{\boldsymbol{\beta}}_b)'\mathbf{X}'\mathbf{X}(\boldsymbol{\beta}_b-\hat{\boldsymbol{\beta}}_b)$$

because of (Koch 1988a, p.60)

$$(\boldsymbol{\beta}_b-\hat{\boldsymbol{\beta}}_b)'\mathbf{X}'(\mathbf{y}-\mathbf{X}\hat{\boldsymbol{\beta}}_b) = (\boldsymbol{\beta}_b-\hat{\boldsymbol{\beta}}_b)'\mathbf{X}'(\mathbf{y}-\mathbf{X}(\mathbf{X}'\mathbf{X})^-_{rs}\mathbf{X}'\mathbf{y}) = 0.$$

The likelihood function is therefore given correspondingly to (224.7) by

$$p(\mathbf{y}|\boldsymbol{\beta}_b,\tau) = (2\pi)^{-n/2}\tau^{n/2}\exp\{-\frac{\tau}{2}[(n-q)\hat{\sigma}^2 + (\boldsymbol{\beta}_b-\hat{\boldsymbol{\beta}}_b)'\mathbf{X}'\mathbf{X}(\boldsymbol{\beta}_b-\hat{\boldsymbol{\beta}}_b)]\}. \tag{315.5}$$

With $h(\mathbf{y})=1$ we see again by the factorization theorem (224.1) that $\hat{\boldsymbol{\beta}}_b$ and $\hat{\sigma}^2$ are sufficient statistics for $\boldsymbol{\beta}$ and σ^2.

a) Noninformative Priors

We will first introduce the noninformative priors (312.2) and (312.4) for the unknown parameters of the model (315.1). Applying Bayes' theorem (211.1) the posterior density function of the parameters $\boldsymbol{\beta}_b$ and τ follows with (315.5) after omitting the constants by

$$p(\boldsymbol{\beta}_b,\tau|\mathbf{y}) \propto \tau^{n/2-1}\exp\{-\frac{\tau}{2}[(n-q)\hat{\sigma}^2 + (\boldsymbol{\beta}_b-\hat{\boldsymbol{\beta}}_b)'\mathbf{X}'\mathbf{X}(\boldsymbol{\beta}_b-\hat{\boldsymbol{\beta}}_b)]\}. \tag{315.6}$$

Like the density function (312.5), this density is proportional to the density of the normal-gamma distribution (A23.1) with the inverse of $\mathbf{X}'\mathbf{X}$ as a parameter. However, this inverse does not exist, so that the normalization constant for (315.6) is equal to zero because of $\det(\mathbf{X}'\mathbf{X})=0$. We therefore use a symmetrical reflexive generalized inverse $(\mathbf{X}'\mathbf{X})^-_{rs}$ of $\mathbf{X}'\mathbf{X}$, which was already introduced for (315.2). If we select q components from $\boldsymbol{\beta}_b$ and $\hat{\boldsymbol{\beta}}_b$ and the corresponding elements from $\mathbf{X}'\mathbf{X}$ such that the resulting matrix becomes regular because of rank$\mathbf{X}'\mathbf{X}$=q, then an inverse $(\mathbf{X}'\mathbf{X})^-_{rs}$ may be computed whose corresponding elements also form a regular matrix (Koch 1988a, p.68). The normal-gamma distribution can then be normalized. Correspondingly to (312.23) we define

$$\boldsymbol{\beta}_b = (\beta_i),\ \boldsymbol{\beta}_{b,j..k} = (\beta_l) \quad \text{and} \quad \hat{\boldsymbol{\beta}}_b = (\hat{\beta}_i),\ \hat{\boldsymbol{\beta}}_{b,j..k} = (\hat{\beta}_l)$$

$$\text{for} \quad l\in\{j,j+1,\ldots,k\},$$

$$\boldsymbol{\Sigma} = (\sigma_{ij}) = (\mathbf{X}'\mathbf{X})^-_{rs} \quad \text{and} \quad \boldsymbol{\Sigma}_{j..k} = (\sigma_{lm}) \quad \text{for} \quad l,m\in\{j,j+1,\ldots,k\}. \tag{315.7}$$

With k-j+1=q we compare the density (315.6) for $\boldsymbol{\beta}_{b,j..k}$ with (A23.1) and find as the posterior distribution because of n/2-1=q/2+(n-q)/2-1 the normal-gamma distribution

$$\boldsymbol{\beta}_{b,j..k},\tau|\mathbf{y} \sim NG(\hat{\boldsymbol{\beta}}_{b,j..k},\boldsymbol{\Sigma}_{j..k},(n-q)\hat{\sigma}^2/2,(n-q)/2) \quad \text{for} \quad k-j+1 = q. \tag{315.8}$$

The marginal posterior distribution of $\boldsymbol{\beta}_{b,j..k}$ for k-j+1=q is therefore according to (A23.3) the multivariate t-distribution and for k-j+1<q because of (A22.10) also the multivariate t-distribution. Thus,

$$\boldsymbol{\beta}_{b,j..k}|\mathbf{y} \sim t(\hat{\boldsymbol{\beta}}_{b,j..k},\hat{\sigma}^2\boldsymbol{\Sigma}_{j..k},n-q) \quad \text{for} \quad k-j+1 \le q. \tag{315.9}$$

According to the definition (315.7) of the vector $\boldsymbol{\beta}_{b,j..k}$ the Bayes estimate $\hat{\boldsymbol{\beta}}_{bB}$ of the entire vector $\boldsymbol{\beta}_b$ from (231.5) and its covariance matrix $\hat{\boldsymbol{\Sigma}}_\beta$ from (231.7) therefore follow with (315.4) and (A22.7) by

$$\hat{\boldsymbol{\beta}}_{bB} = (\mathbf{X}'\mathbf{X})^-_{rs}\mathbf{X}'\mathbf{y} \tag{315.10}$$

and

$$\boldsymbol{\Sigma}_{\hat{\beta}} = \frac{n-q}{n-q-2}\,\hat{\sigma}^2(\mathbf{X}'\mathbf{X})^-_{rs}. \tag{315.11}$$

The Bayes estimate (315.10) is also the best linear unbiased estimate of the projected parameter $\boldsymbol{\beta}_b$ of the standard statistical techniques, its covariance matrix differs by the factor (n-q)/(n-q-2) (Koch 1988a, p.217).

We now look at a linear function $\mathbf{H}\boldsymbol{\beta}_b$ of the projected parameters $\boldsymbol{\beta}_b$, where $\mathbf{H}$ denotes an r×u matrix. Starting from (315.6), we integrate out the parameter τ with n/2-1=q/2+(n-q)/2-1 and with (A12.2) to obtain

$$p(\boldsymbol{\beta}_b|\mathbf{y}) \propto (n-q+(\boldsymbol{\beta}_b-\hat{\boldsymbol{\beta}}_b)'\mathbf{X}'\mathbf{X}(\boldsymbol{\beta}_b-\hat{\boldsymbol{\beta}}_b)/\hat{\sigma}^2)^{-q/2-(n-q)/2}.$$

This density function is proportional to the density function (A22.1) of the multivariate t-distribution with the inverse of $\mathbf{X}'\mathbf{X}/\hat{\sigma}^2$ as a parameter. This inverse does not exist, so that we use $\hat{\sigma}^2(\mathbf{X}'\mathbf{X})^-$ instead, where $(\mathbf{X}'\mathbf{X})^-$ denotes a generalized inverse of $\mathbf{X}'\mathbf{X}$. If

$$\mathrm{rank}(\mathbf{H}(\mathbf{X}'\mathbf{X})^-_{rs}\mathbf{X}'\mathbf{X}) = r \quad \text{for} \quad r \le q \tag{315.12}$$

is fulfilled, then the matrix

$$\mathbf{H}(\mathbf{X}'\mathbf{X})^-_{rs}\mathbf{X}'\mathbf{X}(\mathbf{X}'\mathbf{X})^-\mathbf{X}'\mathbf{X}(\mathbf{X}'\mathbf{X})^-_{rs}\mathbf{H}' = \mathbf{H}(\mathbf{X}'\mathbf{X})^-_{rs}\mathbf{H}'$$

is regular for any choice of the generalized inverse $(\mathbf{X}'\mathbf{X})^-$ (Koch 1988a, p.228). By choosing a regular inverse for the density we therefore obtain for $\mathbf{H}\boldsymbol{\beta}_b$ from (A22.12) the multivariate t-distribution

$$\mathbf{H}\boldsymbol{\beta}_b|\mathbf{y} \sim t(\mathbf{H}\hat{\boldsymbol{\beta}}_b, \hat{\sigma}^2\mathbf{H}(\mathbf{X}'\mathbf{X})^-_{rs}\mathbf{H}', n-q) \quad \text{for} \quad r \le q. \tag{315.13}$$

Furthermore, with (A22.13) we find the distribution, which corresponds to (312.28),

$$(\mathbf{H}\boldsymbol{\beta}_b-\mathbf{H}\hat{\boldsymbol{\beta}}_b)'(\mathbf{H}(\mathbf{X}'\mathbf{X})^-_{rs}\mathbf{H}')^{-1}(\mathbf{H}\boldsymbol{\beta}_b-\mathbf{H}\hat{\boldsymbol{\beta}}_b)/(r\hat{\sigma}^2) \sim F(r,n-q). \tag{315.14}$$

Thus, when establishing confidence regions or when testing the hypothesis

$$H_o : \mathbf{H}\boldsymbol{\beta}_b = \mathbf{w} \quad \text{versus} \quad H_1 : \mathbf{H}\boldsymbol{\beta}_b \neq \mathbf{w}, \tag{315.15}$$

Bayesian inference in the case of noninformative priors gives also for linear models not of full rank results which are equivalent to the standard statistical techniques.

Because of (315.8) the marginal posterior distribution of the weight parameter τ is according to (A23.4) the gamma distribution

$$\tau|\mathbf{y} \sim G((n-q)\hat{\sigma}^2/2, (n-q)/2). \tag{315.16}$$

The Bayes estimate $\hat{\sigma}^2_B$ of σ^2 and its variance thus follow from (312.14) and (312.15) with $n-u$ replaced by $n-q$.

b) Informative Priors

Now we start with informative priors for the parameters $\boldsymbol{\beta}$ and σ^2 of the linear model (315.1) not of full rank. We will assume as in Section 313 that the prior information is introduced by the normal-gamma distribution (313.1)

$$\boldsymbol{\beta}, \tau \sim NG(\boldsymbol{\mu}, \mathbf{V}, b, p) \tag{315.17}$$

with $\tau=1/\sigma^2$ and the parameters of the distribution determined by (313.2).

If the matrix $\mathbf{V}$ is positive definite and therefore regular, then the matrix $\mathbf{V}_o$ in (313.3)

$$\mathbf{V}_o = (\mathbf{X}'\mathbf{X}+\mathbf{V}^{-1})^{-1} \tag{315.18}$$

is also regular, although $\mathbf{X}'\mathbf{X}$ is singular. This is due to the fact that $\mathbf{X}'\mathbf{X}$ is positive semidefinite. For such a case there is no difference to the model of full rank.

Let the matrix $\mathbf{V}$ now be singular. To interpret such a situation, we recapitulate that in the model of full rank the prior information on the parameter vector $\boldsymbol{\beta}$ was introduced according to (313.2) by the vector $\boldsymbol{\mu}_p$ of expected values and by the covariance matrix $\boldsymbol{\Sigma}_\beta$. Solving for $\boldsymbol{\Sigma}_\beta$ we find

$$\boldsymbol{\Sigma}_\beta = \sigma_p^2\mathbf{V} = \sigma_p^2(\mathbf{X}_p'\mathbf{X}_p)^{-1}$$

with

$$\mathbf{V}^{-1} = \mathbf{X}_p'\mathbf{X}_p . \tag{315.19}$$

The inverse $\mathbf{V}^{-1}$ of the matrix $\mathbf{V}$ representing the prior information can therefore be interpreted as originating from prior information $\mathbf{X}_p'\mathbf{X}_p$ on the matrix $\mathbf{X}'\mathbf{X}$ of the normal equations for $\boldsymbol{\beta}$. Thus, if in a model not of full rank the matrix $\mathbf{V}$ is singular, we introduce the prior information by the matrix $\mathbf{X}_p'\mathbf{X}_p$, which represents the prior information on the matrix of normal equations. Furthermore let

$$\operatorname{rank}(\mathbf{X}_p'\mathbf{X}_p) = q \quad \text{and} \quad \operatorname{rank}(\mathbf{X}'\mathbf{X}+\mathbf{X}_p'\mathbf{X}_p) = q . \tag{315.20}$$

The normal-gamma distribution (315.17) cannot be normed, if the matrix $\mathbf{V}$ is singular, but with the same reasoning, which leads from (315.6) to (315.10) and (315.11), and with

$$\mathbf{V}_o = (\mathbf{X}'\mathbf{X}+\mathbf{X}_p'\mathbf{X}_p)_{rs}^-$$

we obtain instead of (313.5) the Bayes estimate $\hat{\boldsymbol{\beta}}_{bB}$ of the projected parameter $\boldsymbol{\beta}_b$ with

$$\hat{\boldsymbol{\beta}}_{bB} = \boldsymbol{\mu}_o = (\mathbf{X}'\mathbf{X}+\mathbf{X}_p'\mathbf{X}_p)_{rs}^-(\mathbf{X}'\mathbf{y}+\mathbf{X}_p'\mathbf{X}_p\boldsymbol{\mu}) \tag{315.21}$$

and instead of (313.6) the covariance matrix $\boldsymbol{\Sigma}_{\hat{\beta}}$ of the estimate

$$\boldsymbol{\Sigma}_{\hat{\beta}} = \frac{1}{n+2p-2}\,[2b+(\boldsymbol{\mu}-\boldsymbol{\mu}_o)'\mathbf{X}_p'\mathbf{X}_p(\boldsymbol{\mu}-\boldsymbol{\mu}_o)+(\mathbf{y}-\mathbf{X}\boldsymbol{\mu}_o)'(\mathbf{y}-\mathbf{X}\boldsymbol{\mu}_o)]\,(\mathbf{X}'\mathbf{X}+\mathbf{X}_p'\mathbf{X}_p)_{rs}^- . \tag{315.22}$$

The estimate $\hat{\sigma}_B^2$ of σ^2 and its variance follows from (313.16) and (313.17) with replacing $\mathbf{V}^{-1}$ by $\mathbf{X}_p'\mathbf{X}_p$.

Example 1: The coordinates of points of a geodetic network determined by distance measurements shall be estimated. Since information is lacking on the position of the network within the coordinate system, in which the coordinates of the netpoints have to be computed, one speaks of a free network, whose datum needs to be defined (Koch 1988a, p.219). Let $\boldsymbol{\beta}$ be the vector of the unknown coordinates of the netpoints, $\mathbf{y}$ the vector of the distance measurements and $\mathbf{X}'\mathbf{X}$ the resulting matrix of normal equations. Let the prior information $\boldsymbol{\mu}_p$ on the coordinates of the network be available by the prior measurements $\mathbf{y}_p$. They lead via the coefficient matrix $\mathbf{X}_p$ to the prior information $\mathbf{X}_p'\mathbf{X}_p$ on the matrix of normal equation, so that (315.20) is fulfilled. The matrix $\mathbf{V}$ representing the prior information is therefore singular and the Bayes estimate $\hat{\boldsymbol{\beta}}_{bB}$ of the coordinates $\boldsymbol{\beta}_b$ projected by (315.2) follows from (315.21) with

$$\hat{\boldsymbol{\beta}}_{bB} = (\mathbf{X}'\mathbf{X}+\mathbf{X}_p'\mathbf{X}_p)_{rs}^{-}(\mathbf{X}'\mathbf{y}+\mathbf{X}_p'\mathbf{X}_p\boldsymbol{\mu}) . \tag{315.23}$$

The vector $\boldsymbol{\mu}_p$ of prior information shall be determined by the estimates of the coordinates resulting from $\mathbf{y}_p$. Hence, we obtain with (313.2) and (315.4)

$$\boldsymbol{\mu} = \boldsymbol{\mu}_p = (\mathbf{X}_p'\mathbf{X}_p)_{rs}^{-}\mathbf{X}_p'\mathbf{y}_p . \tag{315.24}$$

By substituting this result in (315.23) we find because of (Koch 1988a, p.60)

$$\mathbf{X}_p'\mathbf{X}_p(\mathbf{X}_p'\mathbf{X}_p)_{rs}^{-}\mathbf{X}_p' = \mathbf{X}_p' \tag{315.25}$$

the estimate $\hat{\boldsymbol{\beta}}_{bB}$ of the projected coordinates $\boldsymbol{\beta}_b$

$$\hat{\boldsymbol{\beta}}_{bB} = (\mathbf{X}'\mathbf{X}+\mathbf{X}_p'\mathbf{X}_p)_{rs}^{-}(\mathbf{X}'\mathbf{y}+\mathbf{X}_p'\mathbf{y}_p) . \tag{315.26}$$

The estimate $\hat{\boldsymbol{\beta}}_{bB}$ does not depend on the datum chosen in (315.24) to compute the prior information $\boldsymbol{\mu}_p$ on the coordinates. This result had to be expected, since the prior information was assumed to come from the prior measurements $\mathbf{y}_p$ and the resulting matrix $\mathbf{X}_p'\mathbf{X}_p$ only, which do not contain information on the datum.

We will now assume that prior information comes from the prior measurements $\mathbf{y}_p$ with the coefficient matrix $\mathbf{X}_p$ together with information on the datum of the network. This information is introduced by means of a $(u-q)\times u$ matrix $\mathbf{B}$ such that

$$\operatorname{rank}(\mathbf{X}_p'\mathbf{X}_p+\mathbf{B}'\mathbf{B}) = u .$$

Then the matrix $\mathbf{V}$ with

$$\mathbf{V}^{-1} = \mathbf{X}_p'\mathbf{X}_p + \mathbf{B}'\mathbf{B} \tag{315.27}$$

is regular. The matrix $\mathbf{B}$ may be obtained from the matrix $\mathbf{E}$ whose rows constitute with $\mathbf{X}_p\mathbf{E}'=\mathbf{0}$ a basis for the null space of $\mathbf{X}_p$. The estimate

$$\hat{\boldsymbol{\beta}}_b = (\mathbf{X}_p'\mathbf{X}_p+\mathbf{B}'\mathbf{B})^{-1}\mathbf{X}_p'\mathbf{y}_p \tag{315.28}$$

is then given in a datum defined by $\mathbf{B}$ (Koch 1988a, p.217).

The prior information $\boldsymbol{\mu}_p$ on the coordinates shall be given by $\hat{\boldsymbol{\beta}}_b$, so that with (313.2)

$$\boldsymbol{\mu} = \boldsymbol{\mu}_p = \hat{\boldsymbol{\beta}}_b . \tag{315.29}$$

With $\mathbf{V}$ defined by (315.27) the results of the model of full rank apply and we obtain from (313.5) the Bayes estimate $\hat{\boldsymbol{\beta}}_B$ of the coordinates $\boldsymbol{\beta}$ of the netpoints

$$\hat{\boldsymbol{\beta}}_B = (\mathbf{X}'\mathbf{X}+\mathbf{X}_p'\mathbf{X}_p+\mathbf{B}'\mathbf{B})^{-1}(\mathbf{X}'\mathbf{y}+\mathbf{X}_p'\mathbf{y}_p) . \tag{315.30}$$

In contrast to the estimate (315.26) the result is now dependent on the datum, which was introduced by the prior information. ∆

316 Model Identification

There are special linear models for which it is not known at the beginning of the data analysis, how many unknown parameters have to be introduced or what the dimension of the matrix of coefficients should be. Examples are the polynomial model, the autoregressive model for time series or the discrete Fourier series, whose coefficients can be interpreted as being obtained by the method of least squares (Stearns 1975, p.15). We will present the first two models and show that they lead to a general problem of identifying a model. This is also true for the third model. Bayes' theorem is then applied to solve that problem.

a) Polynomial Model

The polynomial model is often used to fit a curve to data points. If a curve is given in a plane, where a rectangular (x,y) coordinate system is defined, and if on the curve points P_i with the coordinates (x_i,y_i) are selected such that for given abscissae x_i the ordinates y_i are measured, then the expected value of the measurement y_i given the u unknown parameters β_1 to β_u collected in $\boldsymbol{\beta}$ may be represented by the polynomial

$$E(y_i|\boldsymbol{\beta}) = \beta_1 + x_i\beta_2 + x_i^2\beta_3 + \ldots + x_i^{u-1}\beta_u \quad \text{for} \quad i\in\{1,\ldots,n\}. \tag{316.1}$$

Polynomial models not in the plane but for higher dimensions are similarly constructed (Koch 1988a, p.231).

For polynomial models generally the question arises, up to which degree the polynomial expansion has to be carried. This means that the integer u in addition to $\boldsymbol{\beta}$ is an unknown parameter. With

$$u \in \{1,\ldots,o\}, \tag{316.2}$$

where o denotes the largest possible degree of the polynomial expansion, which is o=n-1, since with o=n there would be no curve fit any more. We define

$$\mathbf{y} = \begin{bmatrix} y_1 \\ y_2 \\ \ldots \\ y_n \end{bmatrix}, \quad \boldsymbol{\beta}_u = \begin{bmatrix} \beta_{1u} \\ \beta_{2u} \\ \ldots \\ \beta_{uu} \end{bmatrix}, \quad \mathbf{X}_u = \begin{bmatrix} x_1^0 & x_1^1 & \ldots & x_1^{u-1} \\ x_2^0 & x_2^1 & \ldots & x_2^{u-1} \\ \ldots & \ldots & \ldots & \ldots \\ x_n^0 & x_n^1 & \ldots & x_n^{u-1} \end{bmatrix} \tag{316.3}$$

and obtain instead of (316.1) by assuming independent observations with identical variances $\sigma^2=\tau^{-1}$

$$E(\mathbf{y}|\boldsymbol{\beta}_u,u) = \mathbf{X}_u\boldsymbol{\beta}_u \quad \text{with} \quad D(\mathbf{y}|\tau) = \tau^{-1}\mathbf{I}. \tag{316.4}$$

This linear model contains in comparison to the linear model (311.1) together with (311.7) the integer u as unknown parameter.

b) Autoregressive Model

Let y_i with $i\in\{1,\ldots,n\}$ represent a time series, let β_j with $j\in\{1,\ldots,u\}$ denote an unknown parameter and e_i with $i\in\{1,\ldots,n\}$ an error, then

$$y_i = y_{i-1}\beta_1 + y_{i-2}\beta_2 + \ldots + y_{i-u}\beta_u - e_i \tag{316.5}$$

is called an autoregressive process (Box and Jenkins 1970, p.51). The values y_o, $y_{-1},\ldots,y_{1-u}$ are initial observations and assumed to be known. Again the integer u with $u\in\{1,\ldots,o\}$ and o=n-1 is unknown. By defining

$$\mathbf{y} = \begin{bmatrix} y_1 \\ y_2 \\ \ldots \\ y_n \end{bmatrix}, \quad \mathbf{e} = \begin{bmatrix} e_1 \\ e_2 \\ \ldots \\ e_n \end{bmatrix}, \quad \boldsymbol{\beta}_u = \begin{bmatrix} \beta_{1u} \\ \beta_{2u} \\ \ldots \\ \beta_{uu} \end{bmatrix}, \quad \mathbf{X}_u = \begin{bmatrix} y_0 & y_{-1} & \ldots & y_{1-u} \\ y_1 & y_0 & \ldots & y_{2-u} \\ \ldots & \ldots & \ldots & \ldots \\ y_{n-1} & y_{n-2} & \ldots & y_{n-u} \end{bmatrix},$$

$$u \in \{1,\ldots,o\} \tag{316.6}$$

and assuming

$$E(\mathbf{e}|\boldsymbol{\beta}_u,u) = \mathbf{0} \quad \text{and} \quad D(\mathbf{y}|\tau) = \tau^{-1}\mathbf{I} \tag{316.7}$$

we obtain the model

$$E(\mathbf{y}|\boldsymbol{\beta}_u,u) = \mathbf{X}_u\boldsymbol{\beta}_u \quad \text{with} \quad D(\mathbf{y}|\tau) = \tau^{-1}\mathbf{I}. \tag{316.8}$$

If stationarity is not assumed, which means that no constraints are introduced for the parameters, then model (316.8) is identical with model (316.4).

We thus have obtained again a linear model, which contains in comparison to the linear model (311.1) the integer u as an additional unknown parameter.

c) Likelihood Function

To make inferences on u we assume the observations of model (316.4) or (316.8) as being normally distributed. Thus,

$$\mathbf{y}|\boldsymbol{\beta}_u,\tau,u \sim N(\mathbf{X}_u\boldsymbol{\beta}_u,\tau^{-1}\mathbf{I}) \tag{316.9}$$

and with (A21.1) we obtain the likelihood function

$$p(\mathbf{y}|\boldsymbol{\beta}_u,\tau,u) \propto \tau^{n/2}\exp[-\frac{\tau}{2}(\mathbf{y}-\mathbf{X}_u\boldsymbol{\beta}_u)'(\mathbf{y}-\mathbf{X}_u\boldsymbol{\beta}_u)]. \tag{316.10}$$

To apply Bayes' theorem we introduce noninformative and informative priors for the unknown parameters $\boldsymbol{\beta}_u$, τ und u.

d) Noninformative Priors

In agreement with (312.2) and (312.4) we assume the following noninformative prior densities for the unknown parameters $\boldsymbol{\beta}_u$, τ and u

$$p(\boldsymbol{\beta}_u) \propto \text{const.}, \quad p(\tau) \propto 1/\tau, \quad p(u) \propto \text{const.} \tag{316.11}$$

If in addition the parameters are independent, we obtain with (316.10) from Bayes' theorem (211.1) the joint posterior density function for $\boldsymbol{\beta}_u$, τ and u

$$p(\boldsymbol{\beta}_u, \tau, u|\mathbf{y}) \propto \tau^{n/2-1}\exp[-\frac{\tau}{2}(\mathbf{y}-\mathbf{X}_u\boldsymbol{\beta}_u)'(\mathbf{y}-\mathbf{X}_u\boldsymbol{\beta}_u)]. \tag{316.12}$$

With completing the square on $\boldsymbol{\beta}_u$ in the exponent we find in analogy to (224.5)

$$(\mathbf{y}-\mathbf{X}_u\boldsymbol{\beta}_u)'(\mathbf{y}-\mathbf{X}_u\boldsymbol{\beta}_u) = (\mathbf{y}-\mathbf{X}_u\hat{\boldsymbol{\beta}}_u)'(\mathbf{y}-\mathbf{X}_u\hat{\boldsymbol{\beta}}_u) + (\boldsymbol{\beta}_u-\hat{\boldsymbol{\beta}}_u)'\mathbf{X}_u'\mathbf{X}_u(\boldsymbol{\beta}_u-\hat{\boldsymbol{\beta}}_u) \tag{316.13}$$

with

$$\hat{\boldsymbol{\beta}}_u = (\mathbf{X}_u'\mathbf{X}_u)^{-1}\mathbf{X}_u'\mathbf{y}.$$

By substituting (316.13) in (316.12) and by integrating with respect to $\boldsymbol{\beta}_u$ we obtain with (A21.2)

$$\int_{-\infty}^{\infty} \ldots \int_{-\infty}^{\infty} \exp[-\frac{\tau}{2}(\boldsymbol{\beta}_u-\hat{\boldsymbol{\beta}}_u)'\mathbf{X}_u'\mathbf{X}_u(\boldsymbol{\beta}_u-\hat{\boldsymbol{\beta}}_u)]d\beta_{1u}\ldots d\beta_{uu}$$
$$= (2\pi)^{u/2}\tau^{-u/2}(\det\mathbf{X}_u'\mathbf{X}_u)^{-1/2}. \tag{316.14}$$

The marginal posterior density of τ and u therefore follows with

$$p(\tau, u|\mathbf{y}) \propto (2\pi)^{u/2}(\det\mathbf{X}_u'\mathbf{X}_u)^{-1/2}\tau^{(n-u)/2-1}\exp[-\frac{\tau}{2}(\mathbf{y}-\mathbf{X}_u\hat{\boldsymbol{\beta}}_u)'(\mathbf{y}-\mathbf{X}_u\hat{\boldsymbol{\beta}}_u)]. \tag{316.15}$$

Now we integrate with respect to τ and obtain with (A12.2)

$$\int_0^{\infty} \tau^{(n-u)/2-1}\exp[-\frac{\tau}{2}(\mathbf{y}-\mathbf{X}_u\hat{\boldsymbol{\beta}}_u)'(\mathbf{y}-\mathbf{X}_u\hat{\boldsymbol{\beta}}_u)]d\tau$$
$$= \Gamma((n-u)/2)[\frac{1}{2}(\mathbf{y}-\mathbf{X}_u\hat{\boldsymbol{\beta}}_u)'(\mathbf{y}-\mathbf{X}_u\hat{\boldsymbol{\beta}}_u)]^{-(n-u)/2}. \tag{316.16}$$

With this result we finally obtain the posterior density of u

$$p(u|\mathbf{y}) \propto \pi^{u/2}\Gamma((n-u)/2)(\det\mathbf{X}_u'\mathbf{X}_u)^{-1/2}((\mathbf{y}-\mathbf{X}_u\hat{\boldsymbol{\beta}}_u)'(\mathbf{y}-\mathbf{X}_u\hat{\boldsymbol{\beta}}_u))^{-(n-u)/2}. \tag{316.17}$$

For each integer u with $u \in \{1, \ldots, o\}$ we may compute the probability $p(u|\mathbf{y})$ from (316.17) and then select the model for which $p(u|\mathbf{y})$ attains a maximum. The density $p(u|\mathbf{y})$ depends on the sum of squares $(\mathbf{y}-\mathbf{X}_u\hat{\boldsymbol{\beta}}_u)'(\mathbf{y}-\mathbf{X}_u\hat{\boldsymbol{\beta}}_u)$ of the residuals. This sum also enters the information criterion AIC and the final prediction error FPE (Akaike 1979) by which the order of an autoregressive model is estimated (Ulrych and

Bishop 1975).

e) Informative Priors

Again we will assume that no information is available on u. This results in the prior

$$p(u) \propto const. \tag{316.18}$$

However, for the parameters $\boldsymbol{\beta}_u$ and τ we introduce the normal-gamma distribution as prior (Broemeling 1985, p.204)

$$\boldsymbol{\beta}_u, \tau \sim NG(\boldsymbol{\mu}, \mathbf{V}, b, p).$$

With the vector $\boldsymbol{\mu}_p$ of expected values and the covariance matrix $\boldsymbol{\Sigma}_\beta$ of $\boldsymbol{\beta}_u$ and the expected value σ_p^2 and the variance V_{σ^2} of $\sigma^2=1/\tau$ given, the parameters of this distribution are determined by (224.20)

$$\boldsymbol{\mu} = \boldsymbol{\mu}_p,\ \mathbf{V} = \boldsymbol{\Sigma}_\beta/\sigma_p^2,\ p = (\sigma_p^2)^2/V_{\sigma^2} + 2,\ b = (p-1)\sigma_p^2.$$

Thus, with (A23.1) we obtain the prior density

$$p(\boldsymbol{\beta}_u, \tau) \propto (2\pi)^{-u/2}(\det\mathbf{V})^{-1/2}\tau^{u/2+p-1}\exp\{-\frac{\tau}{2}[2b+(\boldsymbol{\beta}_u-\boldsymbol{\mu})'\mathbf{V}^{-1}(\boldsymbol{\beta}_u-\boldsymbol{\mu})]\}. \tag{316.19}$$

Bayes' theorem (211.1) together with (316.10), (316.18) and (316.19) gives the joint posterior density of $\boldsymbol{\beta}_u$, τ and u by

$$p(\boldsymbol{\beta}_u, \tau, u|\mathbf{y}) \propto (2\pi)^{-u/2}(\det\mathbf{V})^{-1/2}\tau^{(n+u)/2+p-1}$$
$$\exp\{-\frac{\tau}{2}[2b+(\boldsymbol{\beta}_u-\boldsymbol{\mu})'\mathbf{V}^{-1}(\boldsymbol{\beta}_u-\boldsymbol{\mu})+(\mathbf{y}-\mathbf{X}_u\boldsymbol{\beta}_u)'(\mathbf{y}-\mathbf{X}_u\boldsymbol{\beta}_u)]\}. \tag{316.20}$$

We complete the squares on $\boldsymbol{\beta}_u$ and obtain with the derivations leading to (224.11)

$$p(\boldsymbol{\beta}_u, \tau, u|\mathbf{y}) \propto (2\pi)^{-u/2}(\det\mathbf{V})^{-1/2}\tau^{(n+u)/2+p-1}$$
$$\exp\{-\frac{\tau}{2}[2b+(\boldsymbol{\mu}-\boldsymbol{\mu}_o)'\mathbf{V}^{-1}(\boldsymbol{\mu}-\boldsymbol{\mu}_o) + (\mathbf{y}-\mathbf{X}_u\boldsymbol{\mu}_o)'(\mathbf{y}-\mathbf{X}_u\boldsymbol{\mu}_o)$$
$$+ (\boldsymbol{\beta}_u-\boldsymbol{\mu}_o)'(\mathbf{X}_u'\mathbf{X}_u+\mathbf{V}^{-1})(\boldsymbol{\beta}_u-\boldsymbol{\mu}_o)]\} \tag{316.21}$$

with

$$\boldsymbol{\mu}_o = (\mathbf{X}_u'\mathbf{X}_u+\mathbf{V}^{-1})^{-1}(\mathbf{X}_u'\mathbf{y}+\mathbf{V}^{-1}\boldsymbol{\mu}). \tag{316.22}$$

We integrate with respect to $\boldsymbol{\beta}_u$ and obtain with (A21.2)

$$\int_{-\infty}^{\infty}\ldots\int_{-\infty}^{\infty}\exp[-\frac{\tau}{2}(\boldsymbol{\beta}_u-\boldsymbol{\mu}_o)'(\mathbf{X}_u'\mathbf{X}_u+\mathbf{V}^{-1})(\boldsymbol{\beta}_u-\boldsymbol{\mu}_o)]d\beta_{1u}\ \ldots\ d\beta_{uu}$$
$$= (2\pi)^{u/2}\tau^{-u/2}(\det(\mathbf{X}_u'\mathbf{X}_u+\mathbf{V}^{-1}))^{-1/2}. \tag{316.23}$$

The marginal density $p(\tau,u|\mathbf{y})$ of τ and u therefore follows from (316.21) with

$$p(\tau,u|\mathbf{y}) \propto (\det\mathbf{V}\det(\mathbf{X}_u'\mathbf{X}_u+\mathbf{V}^{-1}))^{-1/2}\tau^{n/2+p-1}$$
$$\exp\{-\frac{\tau}{2}[2b+(\boldsymbol{\mu}-\boldsymbol{\mu}_o)'\mathbf{V}^{-1}(\boldsymbol{\mu}-\boldsymbol{\mu}_o)+(\mathbf{y}-\mathbf{X}_u\boldsymbol{\mu}_o)'(\mathbf{y}-\mathbf{X}_u\boldsymbol{\mu}_o)]\}. \quad (316.24)$$

Finally we integrate with respect to τ and obtain from (A12.2)

$$\int_0^\infty \tau^{n/2+p-1}\exp\{-\frac{\tau}{2}[2b+(\boldsymbol{\mu}-\boldsymbol{\mu}_o)'\mathbf{V}^{-1}(\boldsymbol{\mu}-\boldsymbol{\mu}_o)+(\mathbf{y}-\mathbf{X}_u\boldsymbol{\mu}_o)'(\mathbf{y}-\mathbf{X}_u\boldsymbol{\mu}_o)]\}d\tau$$
$$= \Gamma(n/2+p)\{\frac{1}{2}[2b+(\boldsymbol{\mu}-\boldsymbol{\mu}_o)'\mathbf{V}^{-1}(\boldsymbol{\mu}-\boldsymbol{\mu}_o)+(\mathbf{y}-\mathbf{X}_u\boldsymbol{\mu}_o)'(\mathbf{y}-\mathbf{X}_u\boldsymbol{\mu}_o)]\}^{-(n/2+p)}. \quad (316.25)$$

With this result we find the posterior density function $p(u|\mathbf{y})$ of u

$$p(u|\mathbf{y}) \propto (\det\mathbf{V}\det(\mathbf{X}_u'\mathbf{X}_u+\mathbf{V}^{-1}))^{-1/2}$$
$$[2b+(\boldsymbol{\mu}-\boldsymbol{\mu}_o)'\mathbf{V}^{-1}(\boldsymbol{\mu}-\boldsymbol{\mu}_o)+(\mathbf{y}-\mathbf{X}_u\boldsymbol{\mu}_o)'(\mathbf{y}-\mathbf{X}_u\boldsymbol{\mu}_o)]^{-(n/2+p)}. \quad (316.26)$$

Again $p(u|\mathbf{y})$ depends on the sum of squares $(\mathbf{y}-\mathbf{X}_u\boldsymbol{\mu}_o)'(\mathbf{y}-\mathbf{X}_u\boldsymbol{\mu}_o)$ of the residuals but now computed with the estimate (316.22) of $\boldsymbol{\beta}_u$. The integer u with $u\in\{1,\ldots,o\}$ for the model identification is chosen such that $p(u|\mathbf{y})$ attains a maximum.

We will now assume that prior information is only available for the variance factor $\sigma^2=1/\tau$ by the expected value σ_p^2 and the variance V_{σ^2} of σ^2. We choose the gamma distribution for the weight parameter τ as prior

$$\tau \sim G(b,p) \quad (316.27)$$

and for $\boldsymbol{\beta}_u$ and u in agreement with (316.11) the priors

$$p(\boldsymbol{\beta}_u) \propto \text{const.}, \quad p(u) \propto \text{const.}$$

With τ having the gamma distribution, σ^2 follows from the inverted gamma distribution $\sigma^2 \sim IG(b,p)$ with expected value and variance from (A13.2). The parameters b and p are therefore determined by

$$p = (\sigma_p^2)^2/V_{\sigma^2} + 2, \quad b = (p-1)\sigma_p^2$$

in agreement with (224.20).

With the prior distribution thus defined, the joint posterior density of $\boldsymbol{\beta}_u$, τ and u follows from Bayes' theorem (211.1) with (316.10) and (A12.1) by

$$p(\boldsymbol{\beta}_u,\tau,u|\mathbf{y}) \propto \tau^{n/2+p-1}\exp\{-\frac{\tau}{2}[2b+(\mathbf{y}-\mathbf{X}_u\boldsymbol{\beta}_u)'(\mathbf{y}-\mathbf{X}_u\boldsymbol{\beta}_u)]\}. \quad (316.28)$$

The integration with respect to $\boldsymbol{\beta}_u$ has been solved by (316.13) and (316.14). Thus,

$$p(\tau,u|\mathbf{y}) \propto (2\pi)^{u/2}(\det\mathbf{X}_u'\mathbf{X}_u)^{-1/2}\tau^{(n-u)/2+p-1}$$
$$\exp\{-\frac{\tau}{2}[2b+(\mathbf{y}-\mathbf{X}_u\hat{\boldsymbol{\beta}}_u)'(\mathbf{y}-\mathbf{X}_u\hat{\boldsymbol{\beta}}_u)]\}.$$

The integration with respect to τ similar to (316.25) finally gives the posterior distribution for u

$$p(u|\mathbf{y}) \propto \pi^{u/2}\Gamma((n-u)/2+p)(\det \mathbf{X}_u'\mathbf{X}_u)^{-1/2}$$
$$[2b+(\mathbf{y}-\mathbf{X}_u\hat{\boldsymbol{\beta}}_u)'(\mathbf{y}-\mathbf{X}_u\hat{\boldsymbol{\beta}}_u)]^{-((n-u)/2+p)} . \tag{316.29}$$

Again the integer u is selected such that $p(u|\mathbf{y})$ attains a maximum.

317 Less Sensitive Hypothesis Tests for the Standard Statistical Techniques

As was shown, the decision to accept or reject the hypothesis (315.15)

$$H_o : \mathbf{H}\boldsymbol{\beta}_b = \mathbf{w} \quad \text{versus} \quad H_1 : \mathbf{H}\boldsymbol{\beta}_b \neq \mathbf{w} \tag{317.1}$$

for the linear model not of full rank is based on the same criterion as in the standard statistical tests. The same holds true for the hypothesis (312.26) of the linear model of full rank. This is due to the noninformative priors for the unknown parameters of the linear model.

When applying the standard statistical techniques, there are cases that the information or the assumption which needs to be tested cannot be exactly expressed by the null hypothesis $\mathbf{H}\boldsymbol{\beta}_b=\mathbf{w}$ in (317.1). Instead of the null vector for the difference $\mathbf{H}\boldsymbol{\beta}_b-\mathbf{w}$, small intervals or regions, which enclose the null vector, should be introduced. As an example, the analysis of measurements is mentioned which are taken at two time epochs to detect the movements of man-made constructions or movements of the earth's crust. In such an analysis the hypothesis is tested that for so-called stable points the differences of the coordinates from two time epochs of observations are equal to zero (Koch 1985). Because of the centering errors of the instruments or because of the residual errors, when taking into account the influence of the atmosphere, it would be more realistic to formulate the hypothesis such that small regions are defined for the coordinate differences, in which they can vary. If in such a situation we nevertheless set $\mathbf{H}\boldsymbol{\beta}_b-\mathbf{w}=\mathbf{0}$, the test reacts too sensitively. This means that the null hypothesis is rejected in cases where, based on the knowledge of the situation to be tested, the acceptance would have been expected.

Bayesian analysis presents no problems when testing hypotheses which are formulated by regions for the parameters, cf. (253.5) and (253.6). We will therefore use the Bayesian approach to develop a test for the hypothesis (317.1) which is less sensitive than the standard statistical test. It shall be derived such that it modifies the F-test of the standard statistical techniques (Koch 1984; Riesmeier 1984). The generalization from the univariate to the multivariate model can be found in Koch and Riesmeier (1985).

Let Δ denote the space of the linear combinations $\mathbf{H}\boldsymbol{\beta}_b$ of the unknown projected parameters $\boldsymbol{\beta}_b$ and let Δ_o denote a subspace of Δ, thus $\Delta_o \subset \Delta$. We formulate corresponding to (252.1) the hypothesis

$$H_o : \mathbf{H}\boldsymbol{\beta}_b \in \Delta_o \quad \text{versus} \quad H_1 : \mathbf{H}\boldsymbol{\beta}_b \in \Delta \backslash \Delta_o . \tag{317.2}$$

The null hypothesis is rejected according to (252.3), if

$$P(\mathbf{H}\boldsymbol{\beta}_b \in \Delta_o | \mathbf{y}) > 1-\alpha \tag{317.3}$$

is fulfilled. As in Section 252, we give the subspace Δ_o a special shape defined by means of the point

$$\mathbf{H}\boldsymbol{\beta}_b = \mathbf{w} \tag{317.4}$$

in (317.1) with

$$T = (\mathbf{H}\hat{\boldsymbol{\beta}}_b - \mathbf{w})'(\mathbf{H}(\mathbf{X}'\mathbf{X})^-_{rs}\mathbf{H}')^{-1}(\mathbf{H}\hat{\boldsymbol{\beta}}_b - \mathbf{w})/(r\hat{\sigma}^2), \tag{317.5}$$

such that

$$\Delta_o = \{\mathbf{H}\boldsymbol{\beta}_b : (\mathbf{H}\hat{\boldsymbol{\beta}}_b - \mathbf{H}\boldsymbol{\beta}_b)'(\mathbf{H}(\mathbf{X}'\mathbf{X})^-_{rs}\mathbf{H}')^{-1}(\mathbf{H}\hat{\boldsymbol{\beta}}_b - \mathbf{H}\boldsymbol{\beta}_b)/(r\hat{\sigma}^2) \le T\}. \tag{317.6}$$

The space Δ_o thus defined has the shape of an hyperellipsoid, cf. (242.6).

If

$$T > F_{1-\alpha;r,n-q}, \tag{317.7}$$

where $F_{1-\alpha;r,n-q}$ denotes the upper α-percentage point of the F-distribution with r and n-q degrees of freedom, we obtain instead of (317.3) because of (315.14)

$$P(\mathbf{H}\boldsymbol{\beta}_b \in \Delta_o | \mathbf{y}) = \int_0^T F(r,n-q)dT > 1-\alpha \tag{317.8}$$

and the null hypothesis in (317.2) is rejected.

As a consequence of the definition of the subspace Δ_o by (317.4) to (317.6), the hypothesis (317.2) may be replaced by the hypothesis (317.1). Hence, the null hypothesis of (317.1) is rejected, if (317.7) is fulfilled. This test procedure is equivalent to the standard F-test with T from (317.5) being the test statistic. This has already been mentioned in connection with (312.30) and (315.15).

It was already pointed out that the test of (317.1) by means of (317.7) may react too sensitively. To obtain a less sensitive test, we restrict the parameter space Δ for $\mathbf{H}\boldsymbol{\beta}_b$. Hence, we introduce a subspace Δ_r with $\Delta_r \subset \Delta$ for the parameters, where they are allowed to vary without contributing to the statistical inference. In Δ_r the posterior density function $p(\mathbf{H}\boldsymbol{\beta}_b|\mathbf{y})$ for $\mathbf{H}\boldsymbol{\beta}_b$ is constrained to zero. Due to the constraint, the posterior density

$p(\mathbf{H}\boldsymbol{\beta}_b|\mathbf{y})$ is truncated, which is admissible according to (225.1), if we renormalize the truncated density.

With the subspace Δ_r the parameter space Δ for $\mathbf{H}\boldsymbol{\beta}_b$ is restricted to the complement Δ_r^c of Δ_r by

$$\Delta_r^c = \Delta \backslash \Delta_r . \tag{317.9}$$

Let the posterior density function for $\mathbf{H}\boldsymbol{\beta}_b$ restricted to Δ_r^c be denoted by $p_c(\mathbf{H}\boldsymbol{\beta}_b|\mathbf{y})$. It is renormalized on Δ_r^c by

$$p_c(\mathbf{H}\boldsymbol{\beta}_b|\mathbf{y}) = (1-\int_{\Delta_r} p(\mathbf{H}\boldsymbol{\beta}_b|\mathbf{y})d\mathbf{H}\boldsymbol{\beta}_b)^{-1} p(\mathbf{H}\boldsymbol{\beta}_b|\mathbf{y}) . \tag{317.10}$$

The probability $P_c(\mathbf{H}\boldsymbol{\beta}_b \in \Delta_o|\mathbf{y})$ computed by the truncated posterior density function $p_c(\mathbf{H}\boldsymbol{\beta}_b|\mathbf{y})$ follows from (317.10), since $p_c(\mathbf{H}\boldsymbol{\beta}_b|\mathbf{y})=0$ in Δ_r, with

$$P_c(\mathbf{H}\boldsymbol{\beta}_b \in \Delta_o|\mathbf{y}) = (1-\int_{\Delta_r} p(\mathbf{H}\boldsymbol{\beta}_b|\mathbf{y})d\mathbf{H}\boldsymbol{\beta}_b)^{-1} \int_{\Delta_o \backslash \Delta_r} p(\mathbf{H}\boldsymbol{\beta}_b|\mathbf{y})d\mathbf{H}\boldsymbol{\beta}_b . \tag{317.11}$$

The subspace Δ_r shall now be defined according to (317.4) to (317.6) by the given point $\mathbf{H}\boldsymbol{\beta}_b=\mathbf{w}_r$. It means that Δ_r takes the shape of an hyperellipsoid. Thus, from (317.8)

$$\int_{\Delta_r} p(\mathbf{H}\boldsymbol{\beta}_b|\mathbf{y})d\mathbf{H}\boldsymbol{\beta}_b = \int_o^{T_r} F(r,n-q)dT$$

with

$$T_r = (\mathbf{H}\hat{\boldsymbol{\beta}}_b-\mathbf{w}_r)'(\mathbf{H}(\mathbf{X}'\mathbf{X})^-_{rs}\mathbf{H}')^{-1}(\mathbf{H}\hat{\boldsymbol{\beta}}_b-\mathbf{w}_r)/(r\hat{\sigma}^2) . \tag{317.12}$$

The probability for the less sensitive test now follows with

$$P_c(\mathbf{H}\boldsymbol{\beta}_b \in \Delta_o|\mathbf{y}) = (1-\int_o^{T_r} F(r,n-q)dT)^{-1} (\int_o^{T} F(r,n-q)dT - \int_o^{T_r} F(r,n-q)dT)$$

$$\text{for} \quad T_r < T \tag{317.13}$$

and

$$P_c(\mathbf{H}\boldsymbol{\beta}_b \in \Delta_o|\mathbf{y}) = 0 \quad \text{for} \quad T_r \geq T . \tag{317.14}$$

We conclude with (317.8) that the null hypothesis in (317.1) is rejected by the less sensitive test, if

$$P_c(\mathbf{H}\boldsymbol{\beta}_b \in \Delta_o|\mathbf{y}) > 1-\alpha$$

or

$$\alpha_T < \alpha , \tag{317.15}$$

where

$$\alpha_T = 1-(1-c)^{-1}(\int_0^T F(r,n-q)dT-c) \qquad (317.16)$$

and

$$\alpha_T = 1 \quad \text{for} \quad T_r \geq T, \; c = \int_0^{T_r} F(r,n-q)dT. \qquad (317.17)$$

The boundary between the region of acceptance and rejection is given by (317.15) with $\alpha_T=\alpha$. Thus, with (317.16)

$$(1-c)^{-1}(\int_0^T F(r,n-q)dT-c) = 1-\alpha$$

or

$$\int_0^T F(r,n-q)dT = 1-\alpha+c\alpha.$$

The null hypothesis in (317.1) is therefore rejected, if

$$T > F_{1-\alpha+c\alpha;r,n-q}, \qquad (317.18)$$

where $F_{1-\alpha+c\alpha;r,n-q}$ denotes the upper $(\alpha-c\alpha)$-percentage point of the F-distribution with r and $n-q$ degrees of freedom.

Because of $0\leq c\leq 1$, it is obvious from (317.18) that the test of the hypothesis (317.1) by (317.15) or (317.18) is less sensitive than the test of the standard statistical techniques.

The sensitivity of the test depends on the size of the subspace Δ_r, where the parameters are allowed to move without contributing to the statistical inference. The size of Δ_r is determined by the quantity T_r from (317.12), which in turn is controlled by the given vector $\mathbf{H}\hat{\boldsymbol{\beta}}_b-\mathbf{w}_r$. The values for the components of $\mathbf{H}\hat{\boldsymbol{\beta}}_b-\mathbf{w}_r$ should be chosen such that they are smaller than the standard deviations computed with $\hat{\sigma}^2$ for the differences $\mathbf{H}\hat{\boldsymbol{\beta}}_b-\mathbf{w}$. If larger values are introduced to compute T_r from (317.12), the test by (317.15) or (317.18) becomes too insensitive.

Example 1: We want to test the hypothesis (312.36) of the Example 2 of Section 312 by the less sensitive test. The test statistic T from (317.5) follows with (312.37) by

$$T = \frac{(\hat{\mu}-\mu_o)^2}{\hat{\sigma}^2/n} .$$

With the following values

$$n = 10, \; \hat{\mu} = 15026.2 \text{ cm}, \; \mu_o = 15027.0 \text{ cm}, \; \hat{\sigma}^2 = 1.0 \text{ cm}^2$$

we compute

$$T = 6.40,$$

so that with $\alpha=0.05$ and

$$F_{1-0.05;1,9} = 5.12$$

the null hypothesis in (312.36) has to be rejected because of (312.37) or (317.7) by the standard test. We now introduce the difference

$$\hat{\mu}-\mu_r = 0.2223 \text{ cm}$$

to define the quantity T_r in (317.12), which in turn determines the region to be excluded from the statistical inference. For this example it is the interval

$$\hat{\mu} \pm 0.2223 \text{ cm}.$$

As recommended, the difference $\hat{\mu}-\mu_r$ is smaller than the standard deviation of $\hat{\mu}-\mu_r$, which is $\hat{\sigma}/\sqrt{n}=0.32$ cm. We compute with (317.12)

$$T_r = n(\hat{\mu}-\mu_r)^2 = 0.494$$

and obtain from a table of percentage points of the F-distribution the value of c from (317.17) by

$$c = \int_0^{T_r} F(1,9)dT = 0.5.$$

With $1-\alpha+c\alpha=1-0.05+0.025=1-0.025$ we find the percentage point

$$F_{1-0.025;1,9} = 7.21,$$

so that according to (317.18) the null hypothesis in (312.36) is accepted by the less sensitive test, while it is rejected by the standard test. Δ

318 Linear Dynamic Systems

So far we have considered linear models whose unknown parameters do not change with time. Now we will introduce the parameters of a dynamic system which are functions of time. The differential equations governing the orbit of a satellite for instance constitute such a dynamic system. The state vector, that is the position and the velocity vector, represents the vector of unknown parameters which depend on time. These unknown parameters have to be estimated. In geodesy, for instance, the parameters of the dynamic system describing the inertial navigation need to be estimated (Schwarz 1983).

Let $\boldsymbol{\beta}_k$ be the u×1 vector of unknown parameters, the so-called *state vector*, at the time epoch k. It is linearly transformed to the unknown u×1 state vector $\boldsymbol{\beta}_{k+1}$ at the epoch k+1 by the u×u *transition matrix* $\boldsymbol{\phi}(k+1,k)$, which is known. A u×1 vector $\mathbf{w}_k$ of random disturbances is also added with $E(\mathbf{w}_k)=\mathbf{0}$ and $D(\mathbf{w}_k)=\mathbf{Q}_k$, where the u×u covariance matrix $\mathbf{Q}_k$ is given. If N epochs are considered, we obtain

$$\boldsymbol{\beta}_{k+1} = \boldsymbol{\phi}(k+1,k)\boldsymbol{\beta}_k + \mathbf{w}_k \quad \text{with} \quad E(\mathbf{w}_k) = \mathbf{0},\ D(\mathbf{w}_k) = \mathbf{Q}_k,\ k\in\{1,\ldots,N-1\}. \tag{318.1}$$

This is called a *linear dynamic system*. It is obtained by the integration of the differential equations constituting the system, see for instance (Jazwinski 1970, p.199). If the differential equations are nonlinear, one has to linearize, e. g. (Jazwinski 1970, p.273).

The observations, which contain the information on the unknown state vectors $\boldsymbol{\beta}_k$, shall establish a linear model. We will introduce it in the general form (311.3). First we assume that the variance factor σ^2 is known and later that it is unknown.

a) Variance Factor Known

Let $\mathbf{y}_k$ be the n×1 vector of observations at the time epoch k, which contains information on $\boldsymbol{\beta}_k$, and $\mathbf{P}_k$ the n×n known weight matrix of $\mathbf{y}_k$, where we have absorbed the variance factor σ^2, which as mentioned is assumed as known. Thus, we obtain instead of (311.3)

$$E(\mathbf{y}_k|\boldsymbol{\beta}_k) = \mathbf{X}_k\boldsymbol{\beta}_k \quad \text{with} \quad D(\mathbf{y}_k) = \mathbf{P}_k^{-1}, \tag{318.2}$$

where the n×u matrix $\mathbf{X}_k$ of coefficients is given.

Let the random disturbances $\mathbf{w}_i$ and $\mathbf{w}_j$, the observations $\mathbf{y}_i$ and $\mathbf{y}_j$ and $\mathbf{w}_i$ and $\mathbf{y}_j$ for $i\neq j$ be independent. In addition, let the vectors be normally distributed

$$\mathbf{w}_k \sim N(\mathbf{0},\mathbf{Q}_k) \tag{318.3}$$

and

$$\mathbf{y}_k|\boldsymbol{\beta}_k \sim N(\mathbf{X}_k\boldsymbol{\beta}_k,\mathbf{P}_k^{-1}). \tag{318.4}$$

The latter distribution leads to the likelihood function for the observation vector $\mathbf{y}_k$.

Based on the observation vector $\mathbf{y}_k$ we want to estimate the unknown state vector $\boldsymbol{\beta}_k$ of the epoch k. The prior information results from the estimate for $\boldsymbol{\beta}_{k-1}$ of the previous epoch based on the observation vector $\mathbf{y}_{k-1}$. The estimate for $\boldsymbol{\beta}_{k-1}$ in turn uses prior information based on $\mathbf{y}_{k-2}$ and so on. Hence, we will recursively apply Bayes' theorem according to (212.1).

Let the prior distribution of the state vector $\boldsymbol{\beta}_1$ of epoch 1 be given by the normal distribution

$$\boldsymbol{\beta}_1|\mathbf{y}_o \sim N(\hat{\boldsymbol{\beta}}_{1,o},\boldsymbol{\Sigma}_{1,o}), \tag{318.5}$$

whose parameters, the vector $\hat{\beta}_{1,o}$ and the covariance matrix $\Sigma_{1,o}$, are given by prior information. The first index in $\hat{\beta}_{1,o}$ and $\Sigma_{1,o}$ refers to the epoch of β_1 and the second index to the epoch of the observation vector y_o, from where the prior information stems. At the beginning no observations are available, therefore y_o. In the following, we will omit this vector y_o.

The prior density function $p(\beta_1)$ defined by (318.5) and the likelihood function $p(y_1|\beta_1)$ from (318.4) are now substituted in Bayes' theorem (211.1) to obtain the posterior density $p(\beta_1|y_1)$. This density is normally distributed, as will be shown below, and after applying the transformation (318.1) gives the prior $p(\beta_2)$ for the parameter vector β_2 of epoch 2. This density has the same form as (318.5).

Since we recursively apply Bayes' theorem, we immediately start with epoch k instead of epoch 1, i.e. with the prior distribution

$$\beta_k|y_1,\ldots,y_{k-1} \sim N(\hat{\beta}_{k,k-1},\Sigma_{k,k-1}). \tag{318.6}$$

Bayes' theorem (211.1) together with the likelihood function from (318.4) gives

$$p(\beta_k|y_1,\ldots,y_k) \propto \exp\{-\frac{1}{2}[(\beta_k-\hat{\beta}_{k,k-1})'\Sigma^{-1}_{k,k-1}(\beta_k-\hat{\beta}_{k,k-1}) \\ +(y_k-X_k\beta_k)'P_k(y_k-X_k\beta_k)]\}. \tag{318.7}$$

Completing the squares on β_k leads to

$$\beta_k'(X_k'P_kX_k+\Sigma^{-1}_{k,k-1})\beta_k - 2\beta_k'(X_k'P_ky_k+\Sigma^{-1}_{k,k-1}\hat{\beta}_{k,k-1}) + y_k'P_ky_k \\ + \hat{\beta}'_{k,k-1}\Sigma^{-1}_{k,k-1}\hat{\beta}_{k,k-1} \\ = y_k'P_ky_k + \hat{\beta}'_{k,k-1}\Sigma^{-1}_{k,k-1}\hat{\beta}_{k,k-1} + (\beta_k-\mu_o)'(X_k'P_kX_k+\Sigma^{-1}_{k,k-1})(\beta_k-\mu_o) \\ - \mu_o'(X_k'P_kX_k+\Sigma^{-1}_{k,k-1})\mu_o$$

with

$$\mu_o = (X_k'P_kX_k+\Sigma^{-1}_{k,k-1})^{-1}(X_k'P_ky_k+\Sigma^{-1}_{k,k-1}\hat{\beta}_{k,k-1}).$$

Substituting this result in (318.7) reveals with (A21.1) the normal distribution as posterior distribution for β_k. Thus,

$$\beta_k|y_1,\ldots,y_k \sim N(\mu_o,(X_k'P_kX_k+\Sigma^{-1}_{k,k-1})^{-1}). \tag{318.8}$$

The expected value is chosen as Bayes estimate $\hat{\beta}_{k,k}$ of β_k according to (231.5), so that we obtain

$$\hat{\beta}_{k,k} = \mu_o \tag{318.9}$$

and its covariance matrix denoted by $\Sigma_{k,k}$ according to (231.7)

$$\Sigma_{k,k} = (X_k' P_k X_k + \Sigma_{k,k-1}^{-1})^{-1}. \tag{318.10}$$

By applying the identities (313.9) we finally derive

$$\hat{\beta}_{k,k} = \hat{\beta}_{k,k-1} + F_k(y_k - X_k \hat{\beta}_{k,k-1}) \tag{318.11}$$

with

$$F_k = \Sigma_{k,k-1} X_k' (X_k \Sigma_{k,k-1} X_k' + P_k^{-1})^{-1}$$

and

$$\Sigma_{k,k} = (I - F_k X_k) \Sigma_{k,k-1}, \tag{318.12}$$

so that $\hat{\beta}_{k,k}$ and $\Sigma_{k,k}$ are recursively computed from $\hat{\beta}_{k,k-1}$ and $\Sigma_{k,k-1}$.

To complete the recursive estimation, we still have to show the transformation of the posterior distribution $p(\beta_{k-1}|y_1,\ldots,y_{k-1})$ for β_{k-1} to the prior distribution (318.6) for β_k. In other words, we have to transform $\hat{\beta}_{k-1,k-1}$ to $\hat{\beta}_{k,k-1}$ and $\Sigma_{k-1,k-1}$ to $\Sigma_{k,k-1}$. We obtain with (318.8) to (318.10)

$$\beta_{k-1}|y_1,\ldots,y_{k-1} \sim N(\hat{\beta}_{k-1,k-1}, \Sigma_{k-1,k-1}).$$

We apply the transformation given by the dynamic system (318.1) to β_{k-1} and find with (318.3) by the theorem for the linear transformation of normally distributed random vectors (Koch 1988a, p.142), since β_{k-1} and w_{k-1} are independent, the distribution for β_k

$$\beta_k|y_1,\ldots,y_{k-1} \sim N(\hat{\beta}_{k,k-1}, \Sigma_{k,k-1}) \tag{318.13}$$

with

$$\hat{\beta}_{k,k-1} = \phi(k,k-1)\hat{\beta}_{k-1,k-1} \tag{318.14}$$

and

$$\Sigma_{k,k-1} = \phi(k,k-1)\Sigma_{k-1,k-1}\phi'(k,k-1) + Q_{k-1}. \tag{318.15}$$

The distribution (318.13) is identical with (318.6), so that we could immediately start from epoch k instead of epoch 1.

The recursive estimation (318.11) and (318.12) together with (318.14) and (318.15) is known as the *Kalman-Bucy filter*. It is applied as a forward filter which updates the knowledge about the state vector whenever new observations are available. These updates may be computed in real time, if the Kalman-Bucy filter is applied, for instance, in aircraft navigation.

The observation vector y_k does not only contain information on $\beta_k, \beta_{k+1}, \ldots, \beta_N$, but also information on $\beta_{k-1}, \beta_{k-2}, \ldots, \beta_1$. After having determined the estimates

$\hat{\boldsymbol{\beta}}_{1,1}, \hat{\boldsymbol{\beta}}_{2,2}, \ldots, \hat{\boldsymbol{\beta}}_{N,N}$ we may therefore go backwards and improve the estimates by introducing the dynamic system (318.1) as observation equation. This backward filter is called smoothing, see for instance (Koch 1982).

b) Variance Factor Unknown

We will now assume that the covariance matrix of the vector $\mathbf{w}_k$ of random disturbances and the covariance matrix of the vector $\mathbf{y}_k$ of observations are known except for the unknown variance factor σ^2. We therefore substitute in (318.1)

$$D(\mathbf{w}_k) = \sigma^2 \mathbf{Q}_k \tag{318.16}$$

and in (318.2)

$$D(\mathbf{y}_k) = \sigma^2 \mathbf{P}_k^{-1} . \tag{318.17}$$

As in the linear model we will work with the weight parameter τ

$$\tau = 1/\sigma^2 \tag{318.18}$$

and obtain instead of (318.3) the normal distribution for $\mathbf{w}_k$ under the condition that τ is given

$$\mathbf{w}_k | \tau \sim N(\mathbf{0}, \tau^{-1}\mathbf{Q}_k) \tag{318.19}$$

and instead of (318.4)

$$\mathbf{y}_k | \boldsymbol{\beta}_k, \tau \sim N(\mathbf{X}_k\boldsymbol{\beta}_k, \tau^{-1}\mathbf{P}_k^{-1}) . \tag{318.20}$$

Let the prior distribution of the state vector $\boldsymbol{\beta}_1$ of epoch 1 given τ be defined as in (318.5) by the normal distribution with the parameters $\hat{\boldsymbol{\beta}}_{1,o}$ and $\tau^{-1}\mathbf{V}_{1,o}$

$$\boldsymbol{\beta}_1 | \tau, \mathbf{y}_o \sim N(\hat{\boldsymbol{\beta}}_{1,o}, \tau^{-1}\mathbf{V}_{1,o}) , \tag{318.21}$$

where again the first index in $\hat{\boldsymbol{\beta}}_{1,o}$ and $\mathbf{V}_{1,o}$ refers to the epoch of $\boldsymbol{\beta}_1$ and the second index to the epoch of the observation vector $\mathbf{y}_o$, from where the prior information stems. The vector $\mathbf{y}_o$ indicates that no observations are available at the beginning, $\mathbf{y}_o$ will be omitted later.

Let the prior distribution of the weight parameter τ be defined by the gamma distribution

$$\tau | \mathbf{y}_o \sim G(b_o, p_o) . \tag{318.22}$$

The joint distribution of $\boldsymbol{\beta}_1$ and τ is therefore determined according to (A23.1) by the normal-gamma distribution

$$\boldsymbol{\beta}_1, \tau | \mathbf{y}_o \sim NG(\hat{\boldsymbol{\beta}}_{1,o}, \mathbf{V}_{1,o}, b_o, p_o) . \tag{318.23}$$

If $\boldsymbol{\mu}_{1p}$ denotes the expected value of $\boldsymbol{\beta}_1$, $\boldsymbol{\Sigma}_{1\beta}$ its covariance matrix, σ_p^2 the expected value of the variance factor σ^2 and V_{σ^2} its variance, the parameters of the normal-gamma distribution (318.23) are determined with (224.20) by

$$\hat{\boldsymbol{\beta}}_{1,o} = \boldsymbol{\mu}_{1p},\ \mathbf{V}_{1,o} = \boldsymbol{\Sigma}_{1\beta}/\sigma_p^2,\ p_o = (\sigma_p^2)^2/V_{\sigma^2} + 2,\ b_o = (p_o-1)\sigma_p^2. \quad (318.24)$$

Again we apply Bayes' theorem recursively and immediately start with epoch k instead of epoch 1, i.e. with the prior distribution for $\boldsymbol{\beta}_k$ and τ with the same form as (318.23)

$$\boldsymbol{\beta}_k,\tau|\mathbf{y}_1,\ldots,\mathbf{y}_{k-1} \sim NG(\hat{\boldsymbol{\beta}}_{k,k-1},\mathbf{V}_{k,k-1},b_{k-1},p_{k-1}). \quad (318.25)$$

Bayes' theorem (211.1) together with the likelihood function from (318.20) leads to the posterior distribution

$$\boldsymbol{\beta}_k,\tau|\mathbf{y}_1,\ldots,\mathbf{y}_k \sim NG(\hat{\boldsymbol{\beta}}_{k,k},\mathbf{V}_{k,k},b_k,p_k) \quad (318.26)$$

with

$$\hat{\boldsymbol{\beta}}_{k,k} = \hat{\boldsymbol{\beta}}_{k,k-1} + \mathbf{F}_k(\mathbf{y}_k-\mathbf{X}_k\hat{\boldsymbol{\beta}}_{k,k-1}) \quad (318.27)$$

$$\mathbf{F}_k = \mathbf{V}_{k,k-1}\mathbf{X}_k'(\mathbf{X}_k\mathbf{V}_{k,k-1}\mathbf{X}_k'+\mathbf{P}_k^{-1})^{-1} \quad (318.28)$$

$$\mathbf{V}_{k,k} = (\mathbf{I}-\mathbf{F}_k\mathbf{X}_k)\mathbf{V}_{k,k-1} \quad (318.29)$$

$$\boldsymbol{\Sigma}_{k,k} = \hat{\sigma}_k^2\mathbf{V}_{k,k} \quad (318.30)$$

$$\hat{\sigma}_k^2 = 2b_k/(n+2p_{k-1}-2) \quad (318.31)$$

$$V(\hat{\sigma}_k^2) = 2(\hat{\sigma}_k^2)^2/(n+2p_{k-1}-4) \quad (318.32)$$

$$b_k = (2b_{k-1}+(\hat{\boldsymbol{\beta}}_{k,k-1}-\hat{\boldsymbol{\beta}}_{k,k})'\mathbf{V}_{k,k-1}^{-1}(\hat{\boldsymbol{\beta}}_{k,k-1}-\hat{\boldsymbol{\beta}}_{k,k})$$
$$+ (\mathbf{y}_k-\mathbf{X}_k\hat{\boldsymbol{\beta}}_{k,k})'\mathbf{P}_k(\mathbf{y}_k-\mathbf{X}_k\hat{\boldsymbol{\beta}}_{k,k}))/2 \quad (318.33)$$

$$p_k = (n+2p_{k-1})/2, \quad (318.34)$$

where $\boldsymbol{\Sigma}_{k,k}$ denotes the covariance matrix of the Bayes estimate $\hat{\boldsymbol{\beta}}_{k,k}$ of $\boldsymbol{\beta}_k$, $\hat{\sigma}_k^2$ the Bayes estimate of epoch k for the variance factor σ^2 and $V(\hat{\sigma}_k^2)$ its variance. These results follow from (224.10), (313.7) to (313.10), (313.16) and (313.17).

To complete the recursive estimation, we still have to show the transformation of the posterior distribution $p(\boldsymbol{\beta}_{k-1},\tau|\mathbf{y}_1,\ldots,\mathbf{y}_{k-1})$ for $\boldsymbol{\beta}_{k-1}$ and τ to the prior distribution (318.25) for $\boldsymbol{\beta}_k$ and τ. We obtain with (318.26)

$$\boldsymbol{\beta}_{k-1},\tau|\mathbf{y}_1,\ldots,\mathbf{y}_{k-1} \sim NG(\hat{\boldsymbol{\beta}}_{k-1,k-1},\mathbf{V}_{k-1,k-1},b_{k-1},p_{k-1}).$$

Because of (A23.2) the distribution of $\boldsymbol{\beta}_{k-1}$ given τ follows with

$$\boldsymbol{\beta}_{k-1}|\tau,\mathbf{y}_1,\ldots,\mathbf{y}_{k-1} \sim N(\hat{\boldsymbol{\beta}}_{k-1,k-1},\tau^{-1}\mathbf{V}_{k-1,k-1}). \quad (318.35)$$

We apply the transformation defined by the dynamic system (318.1) to $\boldsymbol{\beta}_{k-1}$ and find with (318.19) correspondingly to the derivation leading to (318.13)

$$\boldsymbol{\beta}_k | \tau, \mathbf{y}_1, \ldots, \mathbf{y}_{k-1} \sim N(\hat{\boldsymbol{\beta}}_{k,k-1}, \tau^{-1}\mathbf{V}_{k,k-1}) \tag{318.36}$$

with

$$\hat{\boldsymbol{\beta}}_{k,k-1} = \boldsymbol{\phi}(k,k-1)\hat{\boldsymbol{\beta}}_{k-1,k-1} \tag{318.37}$$

and

$$\mathbf{V}_{k,k-1} = \boldsymbol{\phi}(k,k-1)\mathbf{V}_{k-1,k-1}\boldsymbol{\phi}'(k,k-1) + \mathbf{Q}_{k-1}. \tag{318.38}$$

Together with the marginal posterior distribution for τ resulting with (A23.4) from (318.26)

$$\tau | \mathbf{y}_1, \ldots, \mathbf{y}_{k-1} \sim G(b_{k-1}, p_{k-1})$$

we obtain because of (A23.1) the prior distribution

$$\boldsymbol{\beta}_k, \tau | \mathbf{y}_1, \ldots, \mathbf{y}_{k-1} \sim NG(\hat{\boldsymbol{\beta}}_{k,k-1}, \mathbf{V}_{k,k-1}, b_{k-1}, p_{k-1}).$$

This distribution is identical with (318.25), so that we could immediately start from epoch k instead of epoch 1.

Thus, the state vector $\boldsymbol{\beta}_k$ of epoch k is recursively estimated by $\hat{\boldsymbol{\beta}}_{k,k}$ from (318.27) together with (318.28), (318.29), (318.37) and (318.38). The covariance matrix $\boldsymbol{\Sigma}_{k,k}$ of the estimate $\hat{\boldsymbol{\beta}}_{k,k}$ follows from (318.30) with the estimate $\hat{\sigma}_k^2$ for epoch k of the variance factor σ^2 from (318.31). This estimate $\hat{\sigma}_k^2$ is continuously updated by the observations $\mathbf{y}_k$ of each epoch, as can be seen from (318.33). The estimate $\hat{\boldsymbol{\beta}}_{k,k}$ of the state vector $\boldsymbol{\beta}_k$ from (318.27) agrees with the estimate (318.11) of the Kalman-Bucy filter, while the covariance matrix $\boldsymbol{\Sigma}_{k,k}$ differs by the estimated variance factor.

The dynamic system (318.1) together with the linear model (318.2) is also known as the dynamic linear model. It is used, particularly with $\boldsymbol{\phi}(k+1,k)=\mathbf{I}$, to forecast time depending phenomena (West and Harrison 1989).

32 Nonlinear Models

321 Definition and Likelihood Function

Nonlinear models in the standard statistical techniques are linearized by using approximate values for the unknown parameters. The corrections to the approximate values become the unknown parameters and the solutions for the corrections are iterated, in order to improve the approximate values (Koch 1988a, p.185).

For certain types of nonlinear models the exact least squares solution can be derived (Teunissen 1987, 1988). An example, which will be treated in the following section, is fitting a straight line to data points. However, no confidence intervals or hypothesis tests can be given for these least squares solutions.

The Bayesian approach, on the other hand, allows statistical inference in nonlinear models. This will be shown in the following. In general, however, numerical techniques will have to be applied to compute the estimates and the confidence regions of the unknown parameters or to test hypotheses.

As for the linear model, we will assume that the expected values of the observations are defined as given functions of the unkown parameters and that the covariance matrix of the observations is assumed as known except for an unknown factor. But now we allow the observations to be nonlinear functions of the unknown parameters. Nevertheless, by introducing the normal distribution for the observations, the likelihood function of the observations is still specified.

Definition: Let $\mathbf{f}(\boldsymbol{\beta})=(f_i(\boldsymbol{\beta}))$ be an $n\times 1$ given vector of nonlinear functions of the $u\times 1$ vector $\boldsymbol{\beta}$ of unknown random parameters, $\mathbf{y}$ an $n\times 1$ random vector of observations, σ^2 the unknown random variance factor or variance of unit weight and $\mathbf{P}$ the $n\times n$ positive definite weight matrix, which is given. Let $E(\mathbf{y}|\boldsymbol{\beta})$ and $D(\mathbf{y}|\sigma^2)$ be the expected value and the covariance matrix of the observation vector $\mathbf{y}$ under the condition that the unknown parameters $\boldsymbol{\beta}$ and σ^2 are given, then

$$E(\mathbf{y}|\boldsymbol{\beta}) = \mathbf{f}(\boldsymbol{\beta}) \quad \text{with} \quad D(\mathbf{y}|\sigma^2) = \sigma^2\mathbf{P}^{-1}$$

is called a *nonlinear model.* (321.1)

We will assume the normal distribution for the observation vector $\mathbf{y}$ under the condition that $\boldsymbol{\beta}$ and σ^2 are given

$$\mathbf{y}|\boldsymbol{\beta},\sigma^2 \sim N(\mathbf{f}(\boldsymbol{\beta}),\sigma^2\mathbf{P}^{-1}). \tag{321.2}$$

The likelihood function, that is the density of the observations $\mathbf{y}$ given the parameters $\boldsymbol{\beta}$ and σ^2, then follows with (A21.1) by

$$p(\mathbf{y}|\boldsymbol{\beta},\sigma^2) = (2\pi)^{-n/2}(\det\sigma^2\mathbf{P}^{-1})^{-1/2}\exp[-\frac{1}{2\sigma^2}(\mathbf{y}-\mathbf{f}(\boldsymbol{\beta}))'\mathbf{P}(\mathbf{y}-\mathbf{f}(\boldsymbol{\beta}))]. \tag{321.3}$$

Depending on the prior information we may use noninformative priors derived by (222.10) or informative priors. If knowledge is available about the expected values and the covariance matrix of the unknown parameters, the normal distribution for the unknown parameters $\boldsymbol{\beta}$ would be well suited for the prior distribution as in the case of the linear model. For the precision parameter $\tau=1/\sigma^2$ the gamma distribution (A12.1) would be one choice. However, the resulting normal-gamma distribution will not be a conjugate prior, since the likelihood function in (224.10) is different from the likelihood function (321.3). Another choice for τ would be the truncated normal distribution (223.8). In both cases the resulting posterior distribution for the unknown parameters is analytically not tractable, so that the numerical techniques discussed in Section 27 have to be applied for the statistical inference.

322 Fit of a Straight Line

As an application of the nonlinear model we will solve the problem of fitting a straight line to data points (x_i,y_i) with $i\in\{1,\ldots n\}$ in a plane. We will first consider as an introduction the following linear model, where the abscissae x_i of the data points are assumed as known, while the ordinates y_i are independently measured with known variances σ^2. Thus,

$$E(y_i|\beta_0,\beta_1) = \beta_0 + x_i\beta_1 \quad \text{and} \quad V(y_i) = \sigma^2,\ C(y_i,y_j) = 0 \quad \text{for} \quad i \neq j$$
$$\text{and} \quad i,j\in\{1,\ldots,n\}. \tag{322.1}$$

If the method of least squares is applied, the parameters β_0 and β_1 are estimated such that the sum of the squares of the vertical distances of the data points (x_i,y_i) to the straight line are minimized, see Fig. 322-1.

If both the abscissae x_i and the ordinates y_i of the data points are independently measured with known variance σ^2, we obtain instead of (322.1) the model

$$E(y_i|\beta_0,\beta_1) = \beta_0 + E(x_i)\beta_1 \quad \text{for} \quad i\in\{1,\ldots,n\}$$

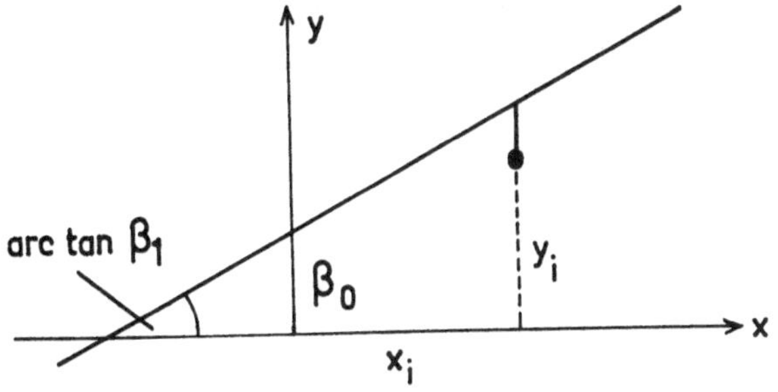

Fig. 322-1

and

$$D(\begin{bmatrix} \mathbf{y} \\ \mathbf{x} \end{bmatrix}) = \sigma^2 \mathbf{I} \quad \text{with} \quad \mathbf{y} = (y_i) \quad \text{and} \quad \mathbf{x} = (x_i). \tag{322.2}$$

By introducing the unknown coordinates μ_i with

$$\mu_i = E(x_i) \quad \text{or} \quad \boldsymbol{\mu} = E(\mathbf{x}) \tag{322.3}$$

the model (322.2) is rewritten by

$$E(\begin{bmatrix} \mathbf{y} \\ \mathbf{x} \end{bmatrix} | \beta_o, \beta_1, \boldsymbol{\mu}) = \begin{bmatrix} \mathbf{e} & \mathbf{I}\beta_1 \\ \mathbf{0} & \mathbf{I} \end{bmatrix} \begin{bmatrix} \beta_o \\ \boldsymbol{\mu} \end{bmatrix} \quad \text{with} \quad D(\begin{bmatrix} \mathbf{y} \\ \mathbf{x} \end{bmatrix}) = \sigma^2 \mathbf{I}, \tag{322.4}$$

where

$$\mathbf{e} = [1, 1, \ldots, 1]'.$$

In contrast to model (322.1), the model (322.4) is nonlinear because of the product $\beta_1 \boldsymbol{\mu}$. Thus, (322.4) is a special case of the nonlinear model (321.1).

By assuming normally distributed observations, the likelihood function of the observations follows with (321.3) by

$$p(\begin{bmatrix} \mathbf{y} \\ \mathbf{x} \end{bmatrix} | \beta_o, \beta_1, \boldsymbol{\mu}) = (2\pi\sigma^2)^{-n/2} \exp[-\frac{1}{2\sigma^2}(\begin{bmatrix} \mathbf{y} \\ \mathbf{x} \end{bmatrix} - \begin{bmatrix} \mathbf{e} & \mathbf{I}\beta_1 \\ \mathbf{0} & \mathbf{I} \end{bmatrix} \begin{bmatrix} \beta_o \\ \boldsymbol{\mu} \end{bmatrix})'$$

$$(\begin{bmatrix} \mathbf{y} \\ \mathbf{x} \end{bmatrix} - \begin{bmatrix} \mathbf{e} & \mathbf{I}\beta_1 \\ \mathbf{0} & \mathbf{I} \end{bmatrix} \begin{bmatrix} \beta_o \\ \boldsymbol{\mu} \end{bmatrix})]$$

or after omitting the constants

$$p(\begin{bmatrix} \mathbf{y} \\ \mathbf{x} \end{bmatrix} | \beta_o, \beta_1, \boldsymbol{\mu}) \propto \exp[-\frac{1}{2\sigma^2}((\mathbf{y} - \mathbf{e}\beta_o - \boldsymbol{\mu}\beta_1)'(\mathbf{y} - \mathbf{e}\beta_o - \boldsymbol{\mu}\beta_1) + (\mathbf{x} - \boldsymbol{\mu})'(\mathbf{x} - \boldsymbol{\mu}))]. \tag{322.5}$$

From (322.5) it becomes obvious that the method of least squares or the maximum likelihood method determines the unknown parameters in (322.4) by minimizing the sum of squares of the perpendicular distances of the data points to the straight line, see Fig. 322-2.

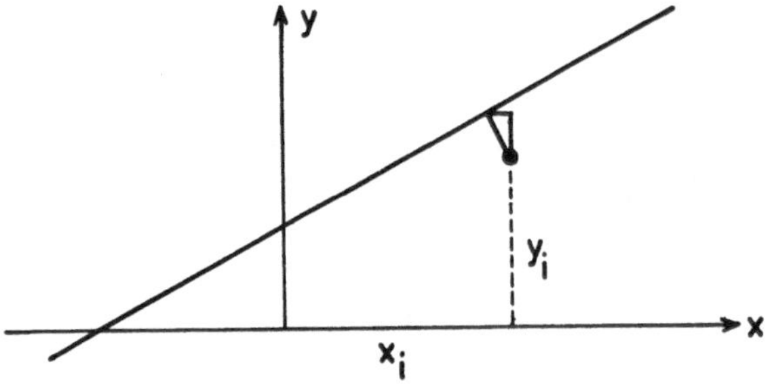

Fig. 322-2

Prior information on the parameters β_o, β_1 and μ_i in (322.4) is assumed as given by means of the expected values and the variances. In addition the parameters are assumed as being independent, so that according to (223.6) the normal distribution is the appropriate distribution. We obtain with

$$E(\beta_o) = \mu_{\beta o},\ V(\beta_o) = V_{\beta o}$$
$$E(\beta_1) = \mu_{\beta 1},\ V(\beta_1) = V_{\beta 1}$$
$$E(\mu_i) = \mu_{oi},\ V(\mu_i) = V_{\mu i} \tag{322.6}$$

and with (A11.1) after omitting the constants the prior density for the parameters β_o, β_1 and $\boldsymbol{\mu}$

$$p(\beta_o, \beta_1, \boldsymbol{\mu}) \propto \exp[-(\beta_o-\mu_{\beta o})^2/2V_{\beta o}]\exp[-(\beta_1-\mu_{\beta 1})^2/2V_{\beta 1}]$$
$$\prod_{i=1}^{n} \exp[-(\mu_i-\mu_{oi})^2/2V_{\mu i}]\,. \tag{322.7}$$

With Bayes' theorem (211.1) and with (322.5) and (322.7) the posterior density function for the unknown parameters β_o, β_1 and $\boldsymbol{\mu}$ follows by

$$p(\beta_o, \beta_1, \boldsymbol{\mu} | \begin{bmatrix} \mathbf{y} \\ \mathbf{x} \end{bmatrix}) \propto \exp\{-\frac{1}{2}[(\beta_o-\mu_{\beta o})^2/V_{\beta o} + (\beta_1-\mu_{\beta 1})^2/V_{\beta 1}$$
$$+ \sum_{i=1}^{n} (\mu_i-\mu_{oi})^2/V_{\mu i} + \sum_{i=1}^{n} (y_i-\beta_o-\mu_i\beta_1)^2/\sigma^2 + \sum_{i=1}^{n} (x_i-\mu_i)^2/\sigma^2]\}\,. \tag{322.8}$$

The parameters β_o and β_1, which determine the straight line, are of special interest. Their marginal distribution is obtained by integrating the parameters μ_i over their domain, that is from $-\infty$ to ∞. The terms in (322.8) containing μ_i are

$$\exp\{-\frac{1}{2}[\sum_{i=1}^{n}((\mu_i-\mu_{oi})^2/V_{\mu i} + (y_i-\beta_o-\mu_i\beta_1)^2/\sigma^2 + (x_i-\mu_i)^2/\sigma^2)]\}$$

$$= \exp\{-\frac{1}{2}[\sum_{i=1}^{n}((\beta_1^2/\sigma^2+1/V_{\mu i}+1/\sigma^2)\mu_i^2-2((y_i-\beta_o)\beta_1/\sigma^2+\mu_{oi}/V_{\mu i}+x_i/\sigma^2)\mu_i$$

$$+ (y_i-\beta_o)^2/\sigma^2 + \mu_{oi}^2/V_{\mu i}+x_i^2/\sigma^2)]\}$$

$$= \prod_{i=1}^{n}\exp[-\frac{1}{2}(\beta_1^2/\sigma^2+1/V_{\mu i}+1/\sigma^2)(\mu_i - \frac{(y_i-\beta_o)\beta_1/\sigma^2+\mu_{oi}/V_{\mu i}+x_i/\sigma^2}{\beta_1^2/\sigma^2+1/V_{\mu i}+1/\sigma^2})^2]$$

$$\exp[-\frac{1}{2}((y_i-\beta_o)^2/\sigma^2+\mu_{oi}^2/V_{\mu i}+x_i^2/\sigma^2 - \frac{((y_i-\beta_o)\beta_1/\sigma^2+\mu_{oi}/V_{\mu i}+x_i/\sigma^2)^2}{\beta_1^2/\sigma^2+1/V_{\mu i}+1/\sigma^2})].$$

The integration over μ_i from $-\infty$ to ∞ leads with (A11.2) after omitting the constant $\sqrt{2\pi}$ to

$$\prod_{i=1}^{n}(\beta_1^2/\sigma^2+1/V_{\mu i}+1/\sigma^2)^{-1/2}\exp[-\frac{1}{2}((y_i-\beta_o)^2/\sigma^2+\mu_{oi}^2/V_{\mu i}+x_i^2/\sigma^2$$

$$- \frac{((y_i-\beta_o)\beta_1/\sigma^2+\mu_{oi}/V_{\mu i}+x_i/\sigma^2)^2}{\beta_1^2/\sigma^2+1/V_{\mu i}+1/\sigma^2})].$$

The marginal posterior density function for the unknown parameters β_o and β_1 thus follows with

$$p(\beta_o,\beta_1|\begin{bmatrix}y\\x\end{bmatrix}) \propto [\prod_{i=1}^{n}(\beta_1^2/\sigma^2+1/V_{\mu i}+1/\sigma^2)^{-1/2}]\exp\{-\frac{1}{2}[(\beta_o-\mu_{\beta o})^2/V_{\beta o}$$

$$+ (\beta_1-\mu_{\beta 1})^2/V_{\beta 1} + \sum_{i=1}^{n}((y_i-\beta_o)^2/\sigma^2+\mu_{oi}^2/V_{\mu i}+x_i^2/\sigma^2$$

$$- \frac{((y_i-\beta_o)\beta_1/\sigma^2+\mu_{oi}/V_{\mu i}+x_i/\sigma^2)^2}{\beta_1^2/\sigma^2+1/V_{\mu i}+1/\sigma^2})]\}. \qquad (322.9)$$

By integrating over β_o the marginal posterior density function for β_1 is obtained. However, the marginal distribution for β_o could not be analytically derived. The numerical techniques discussed in Section 27 should therefore be applied to (322.9) in order to compute the estimates and the confidence intervals for β_o and β_1 or to test hypotheses.

If the prior information given with (322.6) has large variances, so that terms with the variances in the denominator can be neglected, we obtain instead of (322.9) the posterior density function for β_o and β_1 after omitting the constant σ^n

$$p(\beta_o,\beta_1|\begin{bmatrix}\mathbf{y}\\ \mathbf{x}\end{bmatrix}) \propto (\beta_1^2+1)^{-n/2}\exp\{-\frac{1}{2\sigma^2}[\sum_{i=1}^{n}((y_i-\beta_o)^2+x_i^2$$

$$-\frac{((y_i-\beta_o)\beta_1+x_i)^2}{\beta_1^2+1})]\}. \qquad (322.10)$$

Again numerical methods have to be applied for the statistical inference.

Bayes estimates of the parameters β_o and β_1 shall be computed by means of the posterior density function (322.10) for an example. The estimates will be compared with the estimates obtained by applying the method of least squares to the nonlinear model (322.4) (Teunissen 1987). The same results as in Teunissen (1987) are found, if the problem of minimizing the perpendicular distances of data points from a straight line is formulated by the so-called Hessian normal form of a straight line (Wolf 1968, p.419). For easy reference the derivation is given below. It should be pointed out that these methods only give the estimates for the unknown parameters, while the density functions (322.9) or (322.10) allow the estimates and additional statistical inference on the parameters.

The Hessian normal form is given by

$$d_i = x_i\sin\phi - y_i\cos\phi + z, \qquad (322.11)$$

where d_i denotes the perpendicular distance of the data point (x_i, y_i) from the straight line, ϕ the inclination angle of the line and z the perpendicular distance of the origin of the coordinate system from the line, see Fig. 322-3.

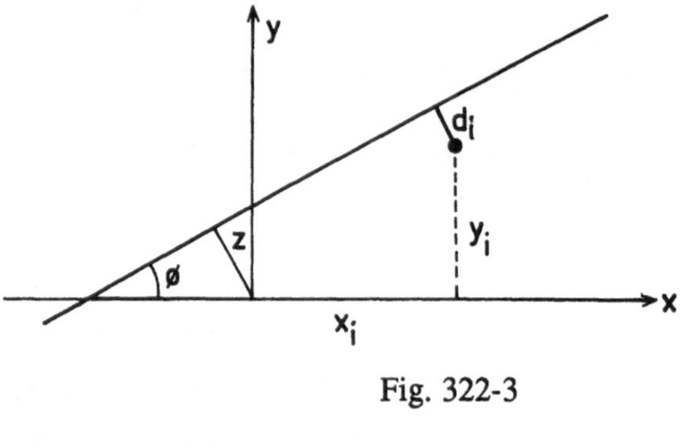

Fig. 322-3

We introduce the coordinates $\bar{x}_i$ and $\bar{y}_i$ referred to the center $x_s = \sum_{i=1}^{n} x_i/n$ and $y_s = \sum_{i=1}^{n} y_i/n$ of the points

$$\bar{x}_i = x_i - x_s,\ \bar{y}_i = y_i - y_s \quad \text{with} \quad \sum_{i=1}^{n}\bar{x}_i = 0,\ \sum_{i=1}^{n}\bar{y}_i = 0. \qquad (322.12)$$

By introducing the vectors

$$\mathbf{d} = (d_i),\ \bar{\mathbf{x}} = (\bar{x}_i),\ \bar{\mathbf{y}} = (\bar{y}_i)$$

the sum $\mathbf{d}'\mathbf{d}$ of the squares of the distances now follows from (322.11) with

$$\mathbf{d}'\mathbf{d} = \bar{\mathbf{x}}'\bar{\mathbf{x}}\ \sin^2\phi + \bar{\mathbf{y}}'\bar{\mathbf{y}}\ \cos^2\phi - 2\bar{\mathbf{x}}'\bar{\mathbf{y}}\ \sin\phi\cos\phi + n\bar{z}^2,$$

where $\bar{z}$ is the value for z with respect to the $\bar{x}_i$, $\bar{y}_i$ coordinates. By setting the derivatives of $\mathbf{d}'\mathbf{d}$ with respect to $\bar{z}$ and ϕ equal to zero

$$\partial\mathbf{d}'\mathbf{d}/\partial\bar{z} = 2n\bar{z} = 0$$

$$\partial\mathbf{d}'\mathbf{d}/\partial\phi = 2\bar{\mathbf{x}}'\bar{\mathbf{x}}\ \sin\phi\cos\phi - 2\bar{\mathbf{y}}'\bar{\mathbf{y}}\ \sin\phi\cos\phi - 2\bar{\mathbf{x}}'\bar{\mathbf{y}}\ \cos2\phi = 0$$

we obtain the estimates of $\bar{z}$ and ϕ

$$\hat{\bar{z}} = 0 \tag{322.13}$$

and

$$\tan2\hat{\phi} = 2\bar{\mathbf{x}}'\bar{\mathbf{y}}/(\bar{\mathbf{x}}'\bar{\mathbf{x}}-\bar{\mathbf{y}}'\bar{\mathbf{y}}). \tag{322.14}$$

Starting from the identity

$$\tan2\phi = 2\tan\phi/(1-\tan^2\phi)$$

we find

$$\tan^2\phi + 2\tan\phi/\tan2\phi - 1 = 0$$

and therefore

$$\tan\phi = (-\ 1 \pm \sqrt{1+\tan^2 2\phi})/\tan2\phi.$$

Now we obtain instead of (322.14)

$$\tan\hat{\phi} = [\bar{\mathbf{y}}'\bar{\mathbf{y}}-\bar{\mathbf{x}}'\bar{\mathbf{x}}\pm((\bar{\mathbf{y}}'\bar{\mathbf{y}}-\bar{\mathbf{x}}'\bar{\mathbf{x}})^2+4(\bar{\mathbf{x}}'\bar{\mathbf{y}})^2)^{1/2}]/2\bar{\mathbf{x}}'\bar{\mathbf{y}}. \tag{322.15}$$

With the estimate $\hat{\beta}_1$ of β_1

$$\hat{\beta}_1 = \tan\hat{\phi} \tag{322.16}$$

and $\hat{\bar{z}}=0$ from (322.13) we have

$$\hat{y}_i - y_s = \hat{\beta}_1(\hat{x}_i - x_s)$$

or

$$\hat{y}_i = (y_s - \hat{\beta}_1 x_s) + \hat{\beta}_1\hat{x}_i.$$

This finally leads to the estimate

$$\hat{\beta}_o = y_s - \hat{\beta}_1 x_s . \tag{322.17}$$

Example 1: For six points the abscissae x_i and the ordinates y_i, given in the following table,

x_i	1.01	1.97	3.00	4.02	4.99	6.02
y_i	0.48	1.01	1.49	2.02	2.51	3.00

have been independently measured with the given variance σ^2=0.0003. Fitting a straight line through these data points by the method of least squares gives with (322.16) and (322.17)

$$\hat{\beta}_o = -0.0055 \quad \text{and} \quad \hat{\beta}_1 = 0.5018 . \tag{322.18}$$

The posterior density function (322.10) was then applied to compute the Bayes estimates $\hat{\beta}_{oB}$ and $\hat{\beta}_{1B}$ of the unknown parameters β_o and β_1 of the straight line in model (322.4) and to determine the confidence interval for the unknown parameter β_o. To obtain this interval for β_o, the marginal posterior density function for β_o was computed from (322.10) by applying (273.2) with i=500 and j=500. This means, for each of the 500 random numbers with uniform distributions generated for β_o, a set of 500 random numbers with uniform distributions was generated for β_1, whose density values were summed according to (273.2). The density values were also multiplied by the random numbers for β_1, in order to compute the Bayes estimate $\hat{\beta}_{1B}$ for β_1 from (272.3).

The marginal posterior density function for β_o was used to compute the Bayes estimate $\hat{\beta}_{oB}$ of β_o from (272.3) and the confidence limits from (272.6). Since the marginal density function is univariate, the random numbers generated for β_o were ordered according to increasing values and along with them the density values. Then ten density values between all consecutive pairs of density values were added by linear interpolation. Finally the confidence limits were computed from (272.6) together with (272.7) by starting from the density values of the minimum and maximum values for β_o and by checking numerically, whether the inequality in (241.2) was fulfilled.

The interval in (271.4) for the coordinate axis of β_1, which defines the domain of the integration, was determined in analogy to (272.1) by finding the density values, which cease to contribute to the integration. The interval for β_o was set up such that the interval of the integration is ten to twenty per cent wider than the confidence interval for β_o, which ensures that only those density values are neglected which do not contribute to the integration.

For each of the two following intervals defining the domain of the integration

$$- 0.06 < \beta_o < 0.05, \quad 0.45 < \beta_1 < 0.54$$

and

$$- 0.05 < \beta_o < 0.04, \quad 0.46 < \beta_1 < 0.53$$

the two sets of 500 times 500 random numbers were generated and $\hat{\beta}_{oB}$, $\hat{\beta}_{1B}$ and the confidence limits for β_o with the content 0.95 were computed. The results are presented in the table (322.19). The Bayes estimates $\hat{\beta}_{oB}$ and $\hat{\beta}_{1B}$ in (322.19) agree very well with the least squares estimates in (322.18).

	$\hat{\beta}_{oB}$	$\hat{\beta}_{1B}$	Confidence Interval
	-0.0045	0.5015	$-0.039 < \beta_o < 0.027$
	-0.0044	0.5016	$-0.041 < \beta_o < 0.029$
	-0.0059	0.5019	$-0.039 < \beta_o < 0.028$
	-0.0057	0.5018	$-0.039 < \beta_o < 0.029$
Mean	-0.0051	0.5017	$-0.040 < \beta_o < 0.028$

(322.19)

Furthermore, the marginal posterior density function of β_o was approximately computed with (274.6) by introducing for β_1 the least squares estimate $\hat{\beta}_1$ from (322.18). For each of the two intervals defining the domain of the integration

$$- 0.04 < \beta_o < 0.03 \quad \text{and} \quad - 0.03 < \beta_o < 0.02$$

two sets of 500 random numbers for β_o were generated, giving the results (322.20).

	$\hat{\beta}_{oB}$	Confidence Interval
	-0.0052	$-0.021 < \beta_o < 0.010$
	-0.0053	$-0.021 < \beta_o < 0.010$
	-0.0059	$-0.021 < \beta_o < 0.010$
	-0.0059	$-0.021 < \beta_o < 0.010$
Mean	-0.0056	$-0.021 < \beta_o < 0.010$

(322.20)

Again the Bayes estimate $\hat{\beta}_{oB}$ agrees very well with the results of (322.18) and (322.19). The confidence interval for β_o is smaller than that of (322.19). It can be explained by the fact that the results of (322.20) are obtained with the parameter β_1 being fixed, while for the results of (322.19) β_1 varies. Δ

33 Mixed Models

331 Mixed Model of the Standard Statistical Techniques

In the standard statistical techniques the mixed model is defined by (Koch 1988a, p.249)

$$\mathbf{X}\boldsymbol{\beta} + \mathbf{Z}\boldsymbol{\gamma} = \mathbf{y} \quad \text{with} \quad E(\boldsymbol{\gamma}) = \mathbf{0} \quad \text{and} \quad D(\boldsymbol{\gamma}) = \sigma^2\boldsymbol{\Sigma}_{\gamma\gamma}, \tag{331.1}$$

where $\mathbf{X}$ and $\mathbf{Z}$ denote the given coefficient matrices, $\boldsymbol{\beta}$ the vector of unknown fixed parameters, $\boldsymbol{\gamma}$ the vector of unknown random parameters, $\mathbf{y}$ the random vector of observations, σ^2 the unknown variance factor and $\boldsymbol{\Sigma}_{\gamma\gamma}$ a given positive definite matrix.

More important than model (331.1) is the following model, which is a special case of (331.1) (Koch 1988a, p.259),

$$\mathbf{X}\boldsymbol{\beta} + \mathbf{Z}\boldsymbol{\gamma} = \mathbf{y} + \mathbf{e} \tag{331.2}$$

with

$$E(\boldsymbol{\gamma}) = \mathbf{0},\ E(\mathbf{e}) = \mathbf{0},\ D(\boldsymbol{\gamma}) = \sigma^2\boldsymbol{\Sigma}_{\gamma\gamma},\ D(\mathbf{e}) = \sigma^2\boldsymbol{\Sigma}_{ee},\ C(\boldsymbol{\gamma},\mathbf{e}) = \mathbf{0}.$$

In addition to (331.1) we have the random vector $\mathbf{e}$, which can be interpreted as error vector of $\mathbf{y}$. The model (331.2) is also called the mixed model or the model of prediction and filtering, since the sum of the observation vector $\mathbf{y}$ and of the error vector $\mathbf{e}$ is represented by the so-called systematic part or the trend $\mathbf{X}\boldsymbol{\beta}$ and the random part $\mathbf{Z}\boldsymbol{\gamma}$, which is called the signal. With the estimates $\hat{\boldsymbol{\beta}}$ and $\hat{\boldsymbol{\gamma}}$ of the unknown parameters $\boldsymbol{\beta}$ and $\boldsymbol{\gamma}$ we may filter the observations by computing $\mathbf{X}\hat{\boldsymbol{\beta}}+\mathbf{Z}\hat{\boldsymbol{\gamma}}$ or predict the observations at points where no observations have been taken by forming $\mathbf{X}^*\hat{\boldsymbol{\beta}}+\mathbf{Z}^*\hat{\boldsymbol{\gamma}}$. The matrices $\mathbf{X}^*$ and $\mathbf{Z}^*$ denote the appropriate coefficient matrices at the points of prediction.

In geodesy, model (331.2) has become well known as the model of least squares collocation (Moritz 1980). It has been successfully applied to interpolate observations or to combine different kinds of data for the determination of the parameters of the gravity field of the earth. If, in addition, coordinates of points at the surface of the earth are determined and if all available observations are used, one arrives at the concept of integrated geodesy (Eeg and Krarup 1973; Hein 1986).

The main problem, when using model (331.2) for the concept of integrated geodesy, is the decision which unknown parameters to associate with the fixed parameters $\boldsymbol{\beta}$ and which with the random parameters $\boldsymbol{\gamma}$. The difference is important, since according to (331.2) the expected value of zero is assumed for $\boldsymbol{\gamma}$. In general, coordinates of points have been connected with $\boldsymbol{\beta}$ and parameters of the gravity field with $\boldsymbol{\gamma}$. The choice might be reasonable for a special application, but it cannot be accepted as a general rule.

While the well-known methods of the standard statistical techniques of testing hypotheses and computing confidence regions for the parameters of linear models can be applied to the fixed parameters $\boldsymbol{\beta}$ of model (331.2), statistical inference for the random parameters $\boldsymbol{\gamma}$ is not well established. Using the approach of the standard statistical tests, hypotheses can be formulated which introduce additional information on the observations $\mathbf{y}$ in (331.2) (Wei 1987). Hypotheses may therefore be tested and confidence regions may be computed for the sum of the trend and the signal.

In the next section we will give a Bayesian interpretation of the mixed model (331.2). Then the question of selecting the parameters according to the property fixed or random does not appear, it is replaced by the question, what prior information exists. In addition there is no difference in the statistical inference for either type of parameters in model (331.2) so that hypotheses for the parameters $\boldsymbol{\gamma}$ may be tested. Hence, the mixed model is much more simply treated from the Bayesian point of view.

332 Definition of the Mixed Model and Likelihood Function

The unknown parameters of the Bayesian analysis are defined as random variables. But, as explained in Section 221, they may represent either fixed quantities or variable quantities. Hence, there is no necessity to distinguish between a fixed parameter $\boldsymbol{\beta}$ and a random parameter $\boldsymbol{\gamma}$ as is done in the mixed models of the standard statistical techniques. The Bayesian analysis distinguishes the parameters according to their prior information.

Looking at model (331.2) under this aspect, it becomes obvious that the parameters $\boldsymbol{\beta}$ and the parameters $\boldsymbol{\gamma}$ differ from what is known about them a priori. No prior information for $\boldsymbol{\beta}$ is available, while the prior information on $\boldsymbol{\gamma}$ covers its expected value and its covariance matrix except for the unknown variance factor. Thus, we obtain the

Definition: Let $\mathbf{X}$ be an $n\times u$ matrix and $\mathbf{Z}$ an $n\times r$ matrix of known coefficients with $\mathrm{rank}\mathbf{X}=u$, $\boldsymbol{\beta}$ a $u\times 1$ vector and $\boldsymbol{\gamma}$ an $r\times 1$ vector of unknown parameters, $\mathbf{y}$ an $n\times 1$ vector of observations, $\mathbf{e}$ an $n\times 1$ vector of errors of the observations with $E(\mathbf{e})=\mathbf{0}$ and $D(\mathbf{e})=\sigma^2\boldsymbol{\Sigma}_{ee}$, where the variance factor σ^2 is unknown and the $n\times n$ matrix $\boldsymbol{\Sigma}_{ee}$ is known and positive definite. Let prior information on the variance factor σ^2 and the parameter vector $\boldsymbol{\gamma}$ be available by $E(\sigma^2)=\sigma_p^2$, $V(\sigma^2)=V_{\sigma^2}$ and by $E(\boldsymbol{\gamma})=\mathbf{0}$, $D(\boldsymbol{\gamma})=\sigma_p^2\boldsymbol{\Sigma}_{\gamma\gamma}$, where the $r\times r$ matrix $\boldsymbol{\Sigma}_{\gamma\gamma}$ is known and positive definite, then

$$\mathbf{X}\boldsymbol{\beta} + \mathbf{Z}\boldsymbol{\gamma} = \mathbf{y} + \mathbf{e}$$

is called a *mixed model.* (332.1)

The expected value and the covariance matrix of the observation vector $\mathbf{y}$ under the condition that the unknown parameters $\boldsymbol{\beta}$, $\boldsymbol{\gamma}$ and σ^2 are given follow with

$$E(\mathbf{y}|\boldsymbol{\beta},\boldsymbol{\gamma}) = \mathbf{X}\boldsymbol{\beta} + \mathbf{Z}\boldsymbol{\gamma} \quad \text{and} \quad D(\mathbf{y}|\boldsymbol{\beta},\boldsymbol{\gamma},\sigma^2) = \sigma^2\boldsymbol{\Sigma}_{ee}, \tag{332.2}$$

where the result for the covariance matrix is derived with $\mathbf{y}=\mathbf{X}\boldsymbol{\beta}+\mathbf{Z}\boldsymbol{\gamma}-\mathbf{e}$ by the law of error propagation (Koch 1988a, p.116). The relation (332.2) is an alternative formulation of the mixed model and is equivalent to the formulation (311.3) of the general linear model. The mixed model (332.1) or (332.2) may therefore be treated as a special linear model, as will be shown in the following sections.

If we assume the observation vector $\mathbf{y}$ as being normally distributed, we obtain with (A21.1) and (A21.3) the likelihood function

$$p(\mathbf{y}|\boldsymbol{\beta},\boldsymbol{\gamma},\sigma^2) = (2\pi)^{-n/2}(\det\sigma^2\boldsymbol{\Sigma}_{ee})^{-1/2}\exp[-\frac{1}{2\sigma^2}(\mathbf{y}-\mathbf{X}\boldsymbol{\beta}-\mathbf{Z}\boldsymbol{\gamma})'\boldsymbol{\Sigma}_{ee}^{-1}(\mathbf{y}-\mathbf{X}\boldsymbol{\beta}-\mathbf{Z}\boldsymbol{\gamma})]. \tag{332.3}$$

333 Posterior Distributions

There is no prior information available for the parameter vector $\boldsymbol{\beta}$ in (332.1), but prior information for $\boldsymbol{\gamma}$ exists by the expected value and the covariance matrix and for σ^2 by the expected value and the variance. We therefore introduce with (224.20) and $\tau=1/\sigma^2$ the normal-gamma distribution (A23.1) as prior distribution. For the parameter vector $\boldsymbol{\beta}$ we choose the null vector as vector of expected values and the null matrix as weight matrix, thus expressing ignorance. After omitting the constants we find

$$p(\begin{bmatrix}\boldsymbol{\beta}\\ \boldsymbol{\gamma}\end{bmatrix},\tau) \propto \tau^{(u+r)/2+p-1}\exp[-\frac{\tau}{2}(2b+\begin{bmatrix}\boldsymbol{\beta}-\mathbf{0}\\ \boldsymbol{\gamma}-\mathbf{0}\end{bmatrix}'\begin{bmatrix}\mathbf{0} & \mathbf{0}\\ \mathbf{0} & \boldsymbol{\Sigma}_{\gamma\gamma}^{-1}\end{bmatrix}\begin{bmatrix}\boldsymbol{\beta}-\mathbf{0}\\ \boldsymbol{\gamma}-\mathbf{0}\end{bmatrix})]. \tag{333.1}$$

This is an improper prior density function, since the matrix of the quadratic form is singular, so that its determinant and the normalization constant for (333.1) are equal to zero.

By applying Bayes' theorem (211.1) we obtain with the prior (333.1) and the likelihood function (332.3) the posterior distribution of $\boldsymbol{\beta}$, $\boldsymbol{\gamma}$ and τ

$$p(\begin{bmatrix}\boldsymbol{\beta}\\ \boldsymbol{\gamma}\end{bmatrix},\tau|\mathbf{y}) \propto \tau^{(u+r)/2+(n+2p)/2-1}\exp[-\frac{\tau}{2}(2b+\begin{bmatrix}\boldsymbol{\beta}-\mathbf{0}\\ \boldsymbol{\gamma}-\mathbf{0}\end{bmatrix}'\begin{bmatrix}\mathbf{0} & \mathbf{0}\\ \mathbf{0} & \boldsymbol{\Sigma}_{\gamma\gamma}^{-1}\end{bmatrix}\begin{bmatrix}\boldsymbol{\beta}-\mathbf{0}\\ \boldsymbol{\gamma}-\mathbf{0}\end{bmatrix})]$$

$$\exp[-\frac{\tau}{2}(\mathbf{y}-[\mathbf{X},\mathbf{Z}]\begin{bmatrix}\boldsymbol{\beta}\\ \boldsymbol{\gamma}\end{bmatrix})'\boldsymbol{\Sigma}_{ee}^{-1}(\mathbf{y}-[\mathbf{X},\mathbf{Z}]\begin{bmatrix}\boldsymbol{\beta}\\ \boldsymbol{\gamma}\end{bmatrix})]. \tag{333.2}$$

We follow the derivation leading to (224.10) and obtain for the exponent of the exponential function

$$2b+\begin{bmatrix}\boldsymbol{\beta}\\ \boldsymbol{\gamma}\end{bmatrix}'\mathbf{V}_o^{-1}\begin{bmatrix}\boldsymbol{\beta}\\ \boldsymbol{\gamma}\end{bmatrix}-2\begin{bmatrix}\boldsymbol{\beta}\\ \boldsymbol{\gamma}\end{bmatrix}'\begin{bmatrix}\mathbf{X}'\boldsymbol{\Sigma}_{ee}^{-1}\mathbf{y}\\ \mathbf{Z}'\boldsymbol{\Sigma}_{ee}^{-1}\mathbf{y}\end{bmatrix}+\mathbf{y}'\boldsymbol{\Sigma}_{ee}^{-1}\mathbf{y}$$

$$=2b+\mathbf{y}'\boldsymbol{\Sigma}_{ee}^{-1}\mathbf{y}+\begin{bmatrix}\boldsymbol{\beta}-\boldsymbol{\beta}_o\\ \boldsymbol{\gamma}-\boldsymbol{\gamma}_o\end{bmatrix}'\mathbf{V}_o^{-1}\begin{bmatrix}\boldsymbol{\beta}-\boldsymbol{\beta}_o\\ \boldsymbol{\gamma}-\boldsymbol{\gamma}_o\end{bmatrix}-\begin{bmatrix}\boldsymbol{\beta}_o\\ \boldsymbol{\gamma}_o\end{bmatrix}'\mathbf{V}_o^{-1}\begin{bmatrix}\boldsymbol{\beta}_o\\ \boldsymbol{\gamma}_o\end{bmatrix}$$

with

$$\mathbf{V}_o=\begin{bmatrix}\mathbf{X}'\boldsymbol{\Sigma}_{ee}^{-1}\mathbf{X} & \mathbf{X}'\boldsymbol{\Sigma}_{ee}^{-1}\mathbf{Z}\\ \mathbf{Z}'\boldsymbol{\Sigma}_{ee}^{-1}\mathbf{X} & \mathbf{Z}'\boldsymbol{\Sigma}_{ee}^{-1}\mathbf{Z}+\boldsymbol{\Sigma}_{\gamma\gamma}^{-1}\end{bmatrix}^{-1}$$

$$\begin{bmatrix}\boldsymbol{\beta}_o\\ \boldsymbol{\gamma}_o\end{bmatrix}=\mathbf{V}_o\begin{bmatrix}\mathbf{X}'\boldsymbol{\Sigma}_{ee}^{-1}\mathbf{y}\\ \mathbf{Z}'\boldsymbol{\Sigma}_{ee}^{-1}\mathbf{y}\end{bmatrix}.$$

Furthermore we have

$$\mathbf{y}'\boldsymbol{\Sigma}_{ee}^{-1}\mathbf{y}-\begin{bmatrix}\boldsymbol{\beta}_o\\ \boldsymbol{\gamma}_o\end{bmatrix}'\mathbf{V}_o^{-1}\begin{bmatrix}\boldsymbol{\beta}_o\\ \boldsymbol{\gamma}_o\end{bmatrix}=\mathbf{y}'\boldsymbol{\Sigma}_{ee}^{-1}\mathbf{y}-2\begin{bmatrix}\boldsymbol{\beta}_o\\ \boldsymbol{\gamma}_o\end{bmatrix}'\begin{bmatrix}\mathbf{X}'\boldsymbol{\Sigma}_{ee}^{-1}\mathbf{y}\\ \mathbf{Z}'\boldsymbol{\Sigma}_{ee}^{-1}\mathbf{y}\end{bmatrix}+\begin{bmatrix}\boldsymbol{\beta}_o\\ \boldsymbol{\gamma}_o\end{bmatrix}'\mathbf{V}_o^{-1}\begin{bmatrix}\boldsymbol{\beta}_o\\ \boldsymbol{\gamma}_o\end{bmatrix}$$

$$=\begin{bmatrix}\boldsymbol{\beta}_o\\ \boldsymbol{\gamma}_o\end{bmatrix}'\begin{bmatrix}\mathbf{0} & \mathbf{0}\\ \mathbf{0} & \boldsymbol{\Sigma}_{\gamma\gamma}^{-1}\end{bmatrix}\begin{bmatrix}\boldsymbol{\beta}_o\\ \boldsymbol{\gamma}_o\end{bmatrix}+(\mathbf{y}-[\mathbf{X},\mathbf{Z}]\begin{bmatrix}\boldsymbol{\beta}_o\\ \boldsymbol{\gamma}_o\end{bmatrix})'\boldsymbol{\Sigma}_{ee}^{-1}(\mathbf{y}-[\mathbf{X},\mathbf{Z}]\begin{bmatrix}\boldsymbol{\beta}_o\\ \boldsymbol{\gamma}_o\end{bmatrix}).$$

This result finally leads to the posterior density

$$p(\begin{bmatrix}\boldsymbol{\beta}\\ \boldsymbol{\gamma}\end{bmatrix},\tau|\mathbf{y}) \propto \tau^{(u+r)/2+(n+2p)/2-1}\exp[-\frac{\tau}{2}(2b+\boldsymbol{\gamma}_o'\boldsymbol{\Sigma}_{\gamma\gamma}^{-1}\boldsymbol{\gamma}_o+(\mathbf{y}-[\mathbf{X},\mathbf{Z}]\begin{bmatrix}\boldsymbol{\beta}_o\\ \boldsymbol{\gamma}_o\end{bmatrix})'$$

$$\boldsymbol{\Sigma}_{ee}^{-1}(\mathbf{y}-[\mathbf{X},\mathbf{Z}]\begin{bmatrix}\boldsymbol{\beta}_o\\ \boldsymbol{\gamma}_o\end{bmatrix})+\begin{bmatrix}\boldsymbol{\beta}-\boldsymbol{\beta}_o\\ \boldsymbol{\gamma}-\boldsymbol{\gamma}_o\end{bmatrix}'\mathbf{V}_o^{-1}\begin{bmatrix}\boldsymbol{\beta}-\boldsymbol{\beta}_o\\ \boldsymbol{\gamma}-\boldsymbol{\gamma}_o\end{bmatrix})].$$

By comparing this density with (A23.1) we recognize that the posterior distribution for $\boldsymbol{\beta}$, $\boldsymbol{\gamma}$ and τ is given by the normal-gamma distribution

$$\begin{bmatrix}\boldsymbol{\beta}\\ \boldsymbol{\gamma}\end{bmatrix},\tau|\mathbf{y} \sim NG(\begin{bmatrix}\boldsymbol{\beta}_o\\ \boldsymbol{\gamma}_o\end{bmatrix},\mathbf{V}_o,b_o,p_o) \tag{333.3}$$

with

$$\mathbf{V}_o = \begin{bmatrix} \mathbf{X}'\boldsymbol{\Sigma}_{ee}^{-1}\mathbf{X} & \mathbf{X}'\boldsymbol{\Sigma}_{ee}^{-1}\mathbf{Z} \\ \mathbf{Z}'\boldsymbol{\Sigma}_{ee}^{-1}\mathbf{X} & \mathbf{Z}'\boldsymbol{\Sigma}_{ee}^{-1}\mathbf{Z}+\boldsymbol{\Sigma}_{\gamma\gamma}^{-1} \end{bmatrix}^{-1}$$

$$\begin{bmatrix} \boldsymbol{\beta}_o \\ \boldsymbol{\gamma}_o \end{bmatrix} = \mathbf{V}_o \begin{bmatrix} \mathbf{X}'\boldsymbol{\Sigma}_{ee}^{-1}\mathbf{y} \\ \mathbf{Z}'\boldsymbol{\Sigma}_{ee}^{-1}\mathbf{y} \end{bmatrix}$$

$$b_o = (2b+\boldsymbol{\gamma}_o'\boldsymbol{\Sigma}_{\gamma\gamma}^{-1}\boldsymbol{\gamma}_o+(\mathbf{y}-\mathbf{X}\boldsymbol{\beta}_o-\mathbf{Z}\boldsymbol{\gamma}_o)'\boldsymbol{\Sigma}_{ee}^{-1}(\mathbf{y}-\mathbf{X}\boldsymbol{\beta}_o-\mathbf{Z}\boldsymbol{\gamma}_o))/2$$

$$p_o = (n+2p)/2$$

and from (224.20)

$$p = (\sigma_p^2)^2/V_{\sigma^2} + 2, \quad b = (p-1)\sigma_p^2 .$$

The marginal posterior distribution for $\boldsymbol{\beta}$ and $\boldsymbol{\gamma}$ follows from (333.3) with (A23.3) by

$$\begin{bmatrix} \boldsymbol{\beta} \\ \boldsymbol{\gamma} \end{bmatrix} | \mathbf{y} \sim t(\begin{bmatrix} \boldsymbol{\beta}_o \\ \boldsymbol{\gamma}_o \end{bmatrix}, b_o\mathbf{V}_o/p_o, 2p_o) . \qquad (333.4)$$

The Bayes estimates $\hat{\boldsymbol{\beta}}_B$ and $\hat{\boldsymbol{\gamma}}_B$ of the unknown parameter vectors $\boldsymbol{\beta}$ and $\boldsymbol{\gamma}$ are computed with (231.5) and (A22.7) by

$$\begin{bmatrix} \hat{\boldsymbol{\beta}}_B \\ \hat{\boldsymbol{\gamma}}_B \end{bmatrix} = \begin{bmatrix} \boldsymbol{\beta}_o \\ \boldsymbol{\gamma}_o \end{bmatrix} = \mathbf{V}_o \begin{bmatrix} \mathbf{X}'\boldsymbol{\Sigma}_{ee}^{-1}\mathbf{y} \\ \mathbf{Z}'\boldsymbol{\Sigma}_{ee}^{-1}\mathbf{y} \end{bmatrix} \qquad (333.5)$$

and the covariance matrix $\boldsymbol{\Sigma}_{\hat{\beta},\hat{\gamma}}$ of the estimates with (231.7) and (A22.7) by

$$\boldsymbol{\Sigma}_{\hat{\beta},\hat{\gamma}} = b_o(p_o-1)^{-1}\mathbf{V}_o . \qquad (333.6)$$

We solve (333.5) for $\hat{\boldsymbol{\gamma}}_B$ and find

$$\hat{\boldsymbol{\gamma}}_B = (\mathbf{Z}'\boldsymbol{\Sigma}_{ee}^{-1}\mathbf{Z}+\boldsymbol{\Sigma}_{\gamma\gamma}^{-1})^{-1}\mathbf{Z}'\boldsymbol{\Sigma}_{ee}^{-1}(\mathbf{y}-\mathbf{X}\hat{\boldsymbol{\beta}}_B) .$$

By applying the first matrix identity of (313.9) we obtain

$$\hat{\boldsymbol{\gamma}}_B = \boldsymbol{\Sigma}_{\gamma\gamma}\mathbf{Z}'(\mathbf{Z}\boldsymbol{\Sigma}_{\gamma\gamma}\mathbf{Z}'+\boldsymbol{\Sigma}_{ee})^{-1}(\mathbf{y}-\mathbf{X}\hat{\boldsymbol{\beta}}_B) , \qquad (333.7)$$

which is identical with the estimate of $\boldsymbol{\gamma}$ in the mixed model (331.2) by the standard statistical techniques (Koch 1988a, p.262). If we substitute the first solution for $\hat{\boldsymbol{\gamma}}_B$ in (333.5) and apply the second matrix identity of (313.9), we find

$$\hat{\boldsymbol{\beta}}_B = (\mathbf{X}'(\mathbf{Z}\boldsymbol{\Sigma}_{\gamma\gamma}\mathbf{Z}'+\boldsymbol{\Sigma}_{ee})^{-1}\mathbf{X})^{-1}\mathbf{X}'(\mathbf{Z}\boldsymbol{\Sigma}_{\gamma\gamma}\mathbf{Z}'+\boldsymbol{\Sigma}_{ee})^{-1}\mathbf{y} , \qquad (333.8)$$

which again is identical with the estimate of $\boldsymbol{\beta}$ by the standard statistical techniques (Koch 1988a, p.263).

The Bayes estimate $\hat{\gamma}_B$ from (333.7) has a form similar to the Bayes estimate $\hat{\beta}_B$ from (313.10). This is due to the fact that the mixed model is a special case of the linear model, as already mentioned in connection with (332.2).

Statistical inference on the individual parameter vectors $\boldsymbol{\beta}$ and $\boldsymbol{\gamma}$ is solved by the marginal distributions resulting from (333.4). We find with (A22.10) the marginal posterior distribution for $\boldsymbol{\beta}$ by

$$\boldsymbol{\beta}|\mathbf{y} \sim t(\boldsymbol{\beta}_o, b_o(\mathbf{X}'\boldsymbol{\Sigma}_{ee}^{-1}\mathbf{X}-\mathbf{X}'\boldsymbol{\Sigma}_{ee}^{-1}\mathbf{Z}(\mathbf{Z}'\boldsymbol{\Sigma}_{ee}^{-1}\mathbf{Z}+\boldsymbol{\Sigma}_{\gamma\gamma}^{-1})^{-1}\mathbf{Z}'\boldsymbol{\Sigma}_{ee}^{-1}\mathbf{X})^{-1}/p_o, 2p_o) . \qquad (333.9)$$

Of special interest is the marginal posterior distribution of $\boldsymbol{\gamma}$

$$\boldsymbol{\gamma}|\mathbf{y} \sim t(\boldsymbol{\gamma}_o, b_o(\mathbf{Z}'\boldsymbol{\Sigma}_{ee}^{-1}\mathbf{Z}+\boldsymbol{\Sigma}_{\gamma\gamma}^{-1}-\mathbf{Z}'\boldsymbol{\Sigma}_{ee}^{-1}\mathbf{X}(\mathbf{X}'\boldsymbol{\Sigma}_{ee}^{-1}\mathbf{X})^{-1}\mathbf{X}'\boldsymbol{\Sigma}_{ee}^{-1}\mathbf{Z})^{-1}/p_o, 2p_o) . \qquad (333.10)$$

All inferential tasks connected with the vector $\boldsymbol{\gamma}$ of unknown parameters can be solved by this distribution.

The marginal posterior distribution of the weight parameter τ is obtained from (333.3) with (A23.4) by

$$\tau \sim G(b_o, p_o) . \qquad (333.11)$$

The posterior distribution for the variance factor σ^2 therefore follows from (A13.1) with

$$\sigma^2 \sim IG(b_o, p_o) . \qquad (333.12)$$

Applying (231.5) and (231.7) gives with (A13.2) the Bayes estimate $\hat{\sigma}_B^2$ of σ^2 and its variance $V(\hat{\sigma}_B^2)$ by

$$\hat{\sigma}_B^2 = b_o/(p_o-1) \quad \text{and} \quad V(\hat{\sigma}_B^2) = b_o^2/((p_o-1)^2(p_o-2)) \qquad (333.13)$$

with b_o and p_o from (333.3).

It was already mentioned in connection with (332.2) that the mixed model may be considered as a special linear model. The posterior distribution (333.3) can therefore be directly obtained by the results of the linear model. This will be shown in the following.

We start from the mixed model (332.2) and introduce bars to differentiate it from the linear model (311.1)

$$E(\bar{\mathbf{y}}|\bar{\boldsymbol{\beta}},\bar{\boldsymbol{\gamma}}) = \bar{\mathbf{X}}\bar{\boldsymbol{\beta}} + \bar{\mathbf{Z}}\bar{\boldsymbol{\gamma}} \quad \text{with} \quad D(\bar{\mathbf{y}}|\bar{\boldsymbol{\beta}},\bar{\boldsymbol{\gamma}},\sigma^2) = \sigma^2\boldsymbol{\Sigma}_{ee} . \qquad (333.14)$$

The matrix $\boldsymbol{\Sigma}_{ee}$ was assumed to be positive definite, so that we apply the Cholesky factorization $\boldsymbol{\Sigma}_{ee}^{-1}=\mathbf{G}\mathbf{G}'$, where $\mathbf{G}$ denotes a regular lower triangular matrix. With

$$\mathbf{X} = \mathbf{G}'[\bar{\mathbf{X}},\bar{\mathbf{Z}}] , \quad \mathbf{y} = \mathbf{G}'\bar{\mathbf{y}} \quad \text{and} \quad \boldsymbol{\beta} = [\bar{\boldsymbol{\beta}}',\bar{\boldsymbol{\gamma}}']' \qquad (333.15)$$

from (311.4) we obtain instead of (333.14) the linear model

$$E(\mathbf{y}|\boldsymbol{\beta}) = \mathbf{X}\boldsymbol{\beta} \quad \text{with} \quad D(\mathbf{y}|\sigma^2) = \sigma^2\mathbf{I},$$

which is identical with (311.1).

The prior density (333.1) for the parameters of the mixed model follows from the prior density (313.1) of the parameters of the linear model by the substitution

$$\boldsymbol{\mu} \rightarrow \begin{bmatrix} \mathbf{0} \\ \mathbf{0} \end{bmatrix}, \quad \mathbf{V}^{-1} \rightarrow \begin{bmatrix} \mathbf{0} & \mathbf{0} \\ \mathbf{0} & \boldsymbol{\Sigma}_{\gamma\gamma}^{-1} \end{bmatrix}. \tag{333.16}$$

By substituting (333.15) and (333.16) in (313.3) we obtain for $\mathbf{V}_o$

$$\mathbf{V}_o = (\bar{\mathbf{X}}'\bar{\mathbf{X}}+\mathbf{V}^{-1})^{-1} = \left(\begin{bmatrix} \bar{\mathbf{X}}' \\ \bar{\mathbf{Z}}' \end{bmatrix} \mathbf{G}\mathbf{G}'[\bar{\mathbf{X}},\bar{\mathbf{Z}}] + \begin{bmatrix} \mathbf{0} & \mathbf{0} \\ \mathbf{0} & \boldsymbol{\Sigma}_{\gamma\gamma}^{-1} \end{bmatrix}\right)^{-1}$$

$$= \begin{bmatrix} \bar{\mathbf{X}}'\boldsymbol{\Sigma}_{ee}^{-1}\bar{\mathbf{X}} & \bar{\mathbf{X}}'\boldsymbol{\Sigma}_{ee}^{-1}\bar{\mathbf{Z}} \\ \bar{\mathbf{Z}}'\boldsymbol{\Sigma}_{ee}^{-1}\bar{\mathbf{X}} & \bar{\mathbf{Z}}'\boldsymbol{\Sigma}_{ee}^{-1}\bar{\mathbf{Z}}+\boldsymbol{\Sigma}_{\gamma\gamma}^{-1} \end{bmatrix}^{-1}$$

and the remaining parameters of the distribution (313.3) accordingly. By omitting the bars we find the distribution (333.3).

334 Prediction and Filtering of Data

As already mentioned in Section 331, the mixed model has been set up for the prediction and filtering of data. If we want to filter observations, we obtain from (332.1) (Koch 1988a, p.260)

$$\mathbf{y}_f = \mathbf{y} + \mathbf{e} = \mathbf{X}\boldsymbol{\beta} + \mathbf{Z}\boldsymbol{\gamma}, \tag{334.1}$$

where the vector $\mathbf{y}_f$ contains the filtered observations. If we want to predict observations, we have

$$\mathbf{y}_p = \mathbf{X}^*\boldsymbol{\beta} + \mathbf{Z}^*\boldsymbol{\gamma}, \tag{334.2}$$

where $\mathbf{X}^*$ and $\mathbf{Z}^*$ denote the coefficient matrices needed to compute the trend and the signal for the predicted observations, which are labeled by $\mathbf{y}_p$. By comparing (334.1) and (334.2) with the alternative formulation (332.2) of the mixed model it becomes obvious that the observations $\mathbf{y}_p$ are predicted as expected values. The distribution for this kind of observations in the linear model is given by (314.2). To derive from it the distribution for $\mathbf{y}_f$ and $\mathbf{y}_p$, we solve $\mathbf{y}_f=\mathbf{G}'\bar{\mathbf{y}}_f$ from (333.15) for $\bar{\mathbf{y}}_f$ and obtain $\bar{\mathbf{y}}_f=\mathbf{G}'^{-1}\mathbf{y}_f$, where $\bar{\mathbf{y}}_f$ now denotes the vector of filtered observations in the mixed model, that is (334.1). Thus, with (314.2), (333.15), (333.16) and (A22.12) we obtain the distributions, where the bars have been omitted,

$$[\mathbf{X},\mathbf{Z}]\begin{bmatrix}\boldsymbol{\beta}\\ \boldsymbol{\gamma}\end{bmatrix}|\mathbf{y} \sim t([\mathbf{X},\mathbf{Z}]\begin{bmatrix}\boldsymbol{\beta}_o\\ \boldsymbol{\gamma}_o\end{bmatrix}, b_o[\mathbf{X},\mathbf{Z}]\mathbf{V}_o\begin{bmatrix}\mathbf{X}'\\ \mathbf{Z}'\end{bmatrix}/p_o, 2p_o) \qquad (334.3)$$

and

$$[\mathbf{X}^*,\mathbf{Z}^*]\begin{bmatrix}\boldsymbol{\beta}\\ \boldsymbol{\gamma}\end{bmatrix}|\mathbf{y} \sim t([\mathbf{X}^*,\mathbf{Z}^*]\begin{bmatrix}\boldsymbol{\beta}_o\\ \boldsymbol{\gamma}_o\end{bmatrix}, b_o[\mathbf{X}^*,\mathbf{Z}^*]\mathbf{V}_o\begin{bmatrix}\mathbf{X}^{*\prime}\\ \mathbf{Z}^{*\prime}\end{bmatrix}/p_o, 2p_o), \qquad (334.4)$$

Of course, the same results can be found from (333.4) with (A22.12).

Instead of the data vector $\mathbf{y}_p$ from (334.2), predicted as an expected value, the actual data vector $\mathbf{y}_u$ shall now be predicted by

$$\mathbf{y}_u = \mathbf{X}^*\boldsymbol{\beta} + \mathbf{Z}^*\boldsymbol{\gamma} - \mathbf{u}, \qquad (334.5)$$

where $\mathbf{u}$ denotes the vector of errors. The predictive distribution for $\mathbf{y}_u$ can be obtained from the predictive distribution (314.21), which was derived for the prediction of unobserved data in the linear model. With $\bar{\mathbf{y}}_u = \mathbf{G}'^{-1}\mathbf{y}_u$ from (333.15), where $\bar{\mathbf{y}}_u$ now denotes the vector of predicted observations in the mixed model, that is (334.5), we find instead of (314.21) with (333.15), (333.16) and (A22.12), if the bars again are omitted,

$$\mathbf{y}_u|\mathbf{y} \sim t([\mathbf{X}^*,\mathbf{Z}^*]\begin{bmatrix}\boldsymbol{\beta}_o\\ \boldsymbol{\gamma}_o\end{bmatrix}, b_o(\boldsymbol{\Sigma}_{ee}+[\mathbf{X}^*,\mathbf{Z}^*]\mathbf{V}_o\begin{bmatrix}\mathbf{X}^{*\prime}\\ \mathbf{Z}^{*\prime}\end{bmatrix})/p_o, 2p_o). \qquad (334.6)$$

The distribution (334.4) for $\mathbf{y}_p$ and the distribution (334.6) for $\mathbf{y}_u$ agree except for the parameter which determines the covariance matrix.

When using the mixed model (331.2) of the standard statistical techniques for the filtering or the prediction, in general not the coefficient matrix $\mathbf{Z}$ itself is given, but the following products with $\mathbf{Z}$ are assumed as known

$$\boldsymbol{\Sigma}_{ss} = \mathbf{Z}\boldsymbol{\Sigma}_{\gamma\gamma}\mathbf{Z}' \qquad (334.7)$$

and

$$\boldsymbol{\Sigma}_{\gamma y} = \boldsymbol{\Sigma}_{\gamma\gamma}\mathbf{Z}' \quad \text{with} \quad \boldsymbol{\Sigma}_{y\gamma} = \mathbf{Z}\boldsymbol{\Sigma}_{\gamma\gamma}. \qquad (334.8)$$

The matrix $\sigma^2\boldsymbol{\Sigma}_{ss}$ is interpreted as the covariance matrix of the signal $\mathbf{Z}\boldsymbol{\gamma}$ and $\sigma^2\boldsymbol{\Sigma}_{\gamma y}$ as the covariance matrix of $\boldsymbol{\gamma}$ and $\mathbf{y}$. In addition, the matrix $\sigma^2\boldsymbol{\Sigma}_{yy}$ with

$$\boldsymbol{\Sigma}_{yy} = \boldsymbol{\Sigma}_{ss} + \boldsymbol{\Sigma}_{ee} \qquad (334.9)$$

is interpreted as the covariance matrix of the observations $\mathbf{y}$.

We substitute (334.7) to (334.9) in (333.7) and (333.8) and obtain

$$\hat{\boldsymbol{\beta}}_B = \boldsymbol{\beta}_o = (\mathbf{X}'\Sigma_{yy}^{-1}\mathbf{X})^{-1}\mathbf{X}'\Sigma_{yy}^{-1}\mathbf{y} \tag{334.10}$$

and

$$\hat{\boldsymbol{\gamma}}_B = \boldsymbol{\gamma}_o = \Sigma_{\gamma y}\Sigma_{yy}^{-1}(\mathbf{y}-\mathbf{X}\hat{\boldsymbol{\beta}}_B). \tag{334.11}$$

These are the well-known estimates of the least squares collocation.

In order to substitute (334.7) to (334.9) in (333.9) and (333.10) we need the following inverses (Koch 1988a, p.40)

$$(\mathbf{Z}'\Sigma_{ee}^{-1}\mathbf{Z}+\Sigma_{\gamma\gamma}^{-1})^{-1} = \Sigma_{\gamma\gamma} - \Sigma_{\gamma\gamma}\mathbf{Z}'(\mathbf{Z}\Sigma_{\gamma\gamma}\mathbf{Z}'+\Sigma_{ee})^{-1}\mathbf{Z}\Sigma_{\gamma\gamma} \tag{334.12}$$

and

$$\begin{aligned}(\mathbf{Z}'\Sigma_{ee}^{-1}\mathbf{Z}+\Sigma_{\gamma\gamma}^{-1}-\mathbf{Z}'\Sigma_{ee}^{-1}\mathbf{X}(\mathbf{X}'\Sigma_{ee}^{-1}\mathbf{X})^{-1}\mathbf{X}'\Sigma_{ee}^{-1}\mathbf{Z})^{-1} &= (\mathbf{Z}'\Sigma_{ee}^{-1}\mathbf{Z}+\Sigma_{\gamma\gamma}^{-1})^{-1}\\ &+ (\mathbf{Z}'\Sigma_{ee}^{-1}\mathbf{Z}+\Sigma_{\gamma\gamma}^{-1})^{-1}\mathbf{Z}'\Sigma_{ee}^{-1}\mathbf{X}(\mathbf{X}'\Sigma_{ee}^{-1}\mathbf{X}-\mathbf{X}'\Sigma_{ee}^{-1}\mathbf{Z}(\mathbf{Z}'\Sigma_{ee}^{-1}\mathbf{Z}+\Sigma_{\gamma\gamma}^{-1})^{-1}\mathbf{Z}'\Sigma_{ee}^{-1}\mathbf{X})^{-1}\\ &\quad \mathbf{X}'\Sigma_{ee}^{-1}\mathbf{Z}(\mathbf{Z}'\Sigma_{ee}^{-1}\mathbf{Z}+\Sigma_{\gamma\gamma}^{-1})^{-1}.\end{aligned} \tag{334.13}$$

With the first inverse we find instead of (333.9) the marginal posterior distribution for $\boldsymbol{\beta}$ by

$$\boldsymbol{\beta}|\mathbf{y} \sim t(\boldsymbol{\beta}_o, b_o(\mathbf{X}'\Sigma_{ee}^{-1}\mathbf{X}-\mathbf{X}'\Sigma_{ee}^{-1}(\Sigma_{ss}-\Sigma_{ss}\Sigma_{yy}^{-1}\Sigma_{ss})\Sigma_{ee}^{-1}\mathbf{X})^{-1}/p_o, 2p_o). \tag{334.14}$$

By substituting (334.12) in (334.13), we find instead of (333.10) the marginal posterior distribution of $\boldsymbol{\gamma}$ with

$$\begin{aligned}\boldsymbol{\gamma}|\mathbf{y} \sim t(\boldsymbol{\gamma}_o, b_o[&\Sigma_{\gamma\gamma}-\Sigma_{\gamma y}\Sigma_{yy}^{-1}\Sigma_{y\gamma}+(\Sigma_{\gamma y}-\Sigma_{\gamma y}\Sigma_{yy}^{-1}\Sigma_{ss})\Sigma_{ee}^{-1}\mathbf{X}(\mathbf{X}'\Sigma_{ee}^{-1}\mathbf{X}-\mathbf{X}'\Sigma_{ee}^{-1}\\ &(\Sigma_{ss}-\Sigma_{ss}\Sigma_{yy}^{-1}\Sigma_{ss})\Sigma_{ee}^{-1}\mathbf{X})^{-1}\mathbf{X}'\Sigma_{ee}^{-1}(\Sigma_{y\gamma}-\Sigma_{ss}\Sigma_{yy}^{-1}\Sigma_{y\gamma})]/p_o, 2p_o).\end{aligned} \tag{334.15}$$

The covariance matrices $D(\hat{\boldsymbol{\beta}})$ and $D(\hat{\boldsymbol{\gamma}})$ of the estimates of the parameter vectors $\boldsymbol{\beta}$ and $\boldsymbol{\gamma}$ in the model (331.2) of the standard statistical techniques are given by (Koch 1988a, p.252,260)

$$D(\hat{\boldsymbol{\beta}}) = \sigma^2(\mathbf{X}'\Sigma_{yy}^{-1}\mathbf{X})^{-1} \tag{334.16}$$

and

$$D(\hat{\boldsymbol{\gamma}}) = \sigma^2(\Sigma_{\gamma y}\Sigma_{yy}^{-1}\Sigma_{y\gamma}-\Sigma_{\gamma y}\Sigma_{yy}^{-1}\mathbf{X}(\mathbf{X}'\Sigma_{yy}^{-1}\mathbf{X})^{-1}\mathbf{X}'\Sigma_{yy}^{-1}\Sigma_{y\gamma}). \tag{334.17}$$

By comparing these covariance matrices with the covariance matrices for the Bayesian estimates $\hat{\boldsymbol{\beta}}_B$ and $\hat{\boldsymbol{\gamma}}_B$, which follow with (231.7) and (A22.7) from (334.14) and (334.15), it becomes obvious that the covariance matrices disagree, although the estimates themselves agree.

335 Special Model for Prediction and Filtering of Data

We substitute in (332.1)

$$\mathbf{s} = \mathbf{Z}\boldsymbol{\gamma}, \tag{335.1}$$

where the n×1 vector $\mathbf{s}$ denotes the signal, so that the model is obtained

$$\mathbf{X}\boldsymbol{\beta} + \mathbf{s} = \mathbf{y} + \mathbf{e}. \tag{335.2}$$

The data $\mathbf{y}$ now directly contains the information on the signal $\mathbf{s}$, while $\mathbf{X}\boldsymbol{\beta}$ again represents the systematic part or the trend of the signal. Thus, (335.2) is a special model for the prediction and filtering of data.

With (335.1) the posterior distribution (334.14) for $\boldsymbol{\beta}$ does not change. However, instead of (334.15), we find with (334.7), (334.8), (334.11) and (A22.12) the posterior distribution for the signal $\mathbf{s}$

$$\mathbf{s}|\mathbf{y} \sim t(\mathbf{s}_o, b_o[\boldsymbol{\Sigma}_{ss} - \boldsymbol{\Sigma}_{ss}\boldsymbol{\Sigma}_{yy}^{-1}\boldsymbol{\Sigma}_{ss} + (\boldsymbol{\Sigma}_{ss} - \boldsymbol{\Sigma}_{ss}\boldsymbol{\Sigma}_{yy}^{-1}\boldsymbol{\Sigma}_{ss})\boldsymbol{\Sigma}_{ee}^{-1}\mathbf{X}(\mathbf{X}'\boldsymbol{\Sigma}_{ee}^{-1}\mathbf{X} - \mathbf{X}'\boldsymbol{\Sigma}_{ee}^{-1}$$

$$(\boldsymbol{\Sigma}_{ss} - \boldsymbol{\Sigma}_{ss}\boldsymbol{\Sigma}_{yy}^{-1}\boldsymbol{\Sigma}_{ss})\boldsymbol{\Sigma}_{ee}^{-1}\mathbf{X})^{-1}\mathbf{X}'\boldsymbol{\Sigma}_{ee}^{-1}(\boldsymbol{\Sigma}_{ss} - \boldsymbol{\Sigma}_{ss}\boldsymbol{\Sigma}_{yy}^{-1}\boldsymbol{\Sigma}_{ss})]/p_o, 2p_o), \tag{335.3}$$

where with $\hat{\mathbf{s}}_B = \mathbf{Z}\hat{\boldsymbol{\gamma}}_B$ and $\mathbf{s}_o = \mathbf{Z}\boldsymbol{\gamma}_o$

$$\hat{\mathbf{s}}_B = \mathbf{s}_o = \boldsymbol{\Sigma}_{ss}\boldsymbol{\Sigma}_{yy}^{-1}(\mathbf{y} - \mathbf{X}\hat{\boldsymbol{\beta}}_B)$$

and with (333.3) because of $\boldsymbol{\gamma}_o'\boldsymbol{\Sigma}_{\gamma\gamma}^{-1}\boldsymbol{\gamma}_o = (\mathbf{y} - \mathbf{X}\boldsymbol{\beta}_o)'\boldsymbol{\Sigma}_{yy}^{-1}\boldsymbol{\Sigma}_{y\gamma}\boldsymbol{\Sigma}_{\gamma\gamma}^{-1}\boldsymbol{\Sigma}_{\gamma y}\boldsymbol{\Sigma}_{yy}^{-1}(\mathbf{y} - \mathbf{X}\boldsymbol{\beta}_o) = (\mathbf{y} - \mathbf{X}\boldsymbol{\beta}_o)'$ $\boldsymbol{\Sigma}_{yy}^{-1}\boldsymbol{\Sigma}_{ss}\boldsymbol{\Sigma}_{yy}^{-1}(\mathbf{y} - \mathbf{X}\boldsymbol{\beta}_o)$ from (334.7), (334.8), (334.10) and (334.11)

$$b_o = (2b + (\mathbf{y} - \mathbf{X}\boldsymbol{\beta}_o)'\boldsymbol{\Sigma}_{yy}^{-1}\boldsymbol{\Sigma}_{ss}\boldsymbol{\Sigma}_{yy}^{-1}(\mathbf{y} - \mathbf{X}\boldsymbol{\beta}_o) + (\mathbf{y} - \mathbf{X}\boldsymbol{\beta}_o - \mathbf{s}_o)'\boldsymbol{\Sigma}_{ee}^{-1}(\mathbf{y} - \mathbf{X}\boldsymbol{\beta}_o - \mathbf{s}_o))/2$$

$$p_o = (n+2p)/2, \ p = (\sigma_p^2)^2/V_{\sigma^2} + 2, \ b = (p-1)\sigma_p^2.$$

All inferential problems connected with the signal $\mathbf{s}$ can be solved by this distribution.

We now derive the distribution of a predicted signal. We substitute in (334.2)

$$\mathbf{s}^* = \mathbf{Z}^*\boldsymbol{\gamma}, \tag{335.4}$$

where the n×1 vector $\mathbf{s}^*$ denotes the predicted signal, and obtain the observations $\mathbf{y}_p$ predicted as expected values

$$\mathbf{y}_p = \mathbf{X}^*\boldsymbol{\beta} + \mathbf{s}^*. \tag{335.5}$$

We introduce

$$\boldsymbol{\Sigma}_{s^*s^*} = \mathbf{Z}^*\boldsymbol{\Sigma}_{\gamma\gamma}\mathbf{Z}^{*\prime}, \ \boldsymbol{\Sigma}_{s^*s} = \mathbf{Z}^*\boldsymbol{\Sigma}_{\gamma\gamma}\mathbf{Z}' \ \text{ and } \ \boldsymbol{\Sigma}_{ss^*} = \mathbf{Z}\,\boldsymbol{\Sigma}_{\gamma\gamma}\mathbf{Z}^{*\prime}, \tag{335.6}$$

where $\boldsymbol{\Sigma}_{s^*s^*}$ is interpreted as the covariance matrix of the predicted signal and $\boldsymbol{\Sigma}_{s^*s}$ the covariance matrix of the predicted and the original signal (Koch 1988a, p.261). With (335.4) and (335.6) we obtain instead of (334.15) the posterior distribution for the predicted signal $\mathbf{s}^*$

$$\mathbf{s}^*|\mathbf{y} \sim t(\mathbf{s}_o^*, b_o[\boldsymbol{\Sigma}_{s^*s^*} - \boldsymbol{\Sigma}_{s^*s}\boldsymbol{\Sigma}_{yy}^{-1}\boldsymbol{\Sigma}_{ss^*} + (\boldsymbol{\Sigma}_{s^*s} - \boldsymbol{\Sigma}_{s^*s}\boldsymbol{\Sigma}_{yy}^{-1}\boldsymbol{\Sigma}_{ss})\boldsymbol{\Sigma}_{ee}^{-1}\mathbf{X}(\mathbf{X}'\boldsymbol{\Sigma}_{ee}^{-1}\mathbf{X} - \mathbf{X}'\boldsymbol{\Sigma}_{ee}^{-1}$$

$$(\boldsymbol{\Sigma}_{ss} - \boldsymbol{\Sigma}_{ss}\boldsymbol{\Sigma}_{yy}^{-1}\boldsymbol{\Sigma}_{ss})\boldsymbol{\Sigma}_{ee}^{-1}\mathbf{X})^{-1}\mathbf{X}'\boldsymbol{\Sigma}_{ee}^{-1}(\boldsymbol{\Sigma}_{ss^*} - \boldsymbol{\Sigma}_{ss}\boldsymbol{\Sigma}_{yy}^{-1}\boldsymbol{\Sigma}_{ss^*})]/p_o, 2p_o), \quad (335.7)$$

where with $\hat{\mathbf{s}}_B^* = \mathbf{Z}^*\hat{\boldsymbol{\gamma}}_B$ and $\mathbf{s}_o^* = \mathbf{Z}^*\boldsymbol{\gamma}_o$

$$\hat{\mathbf{s}}_B^* = \mathbf{s}_o^* = \boldsymbol{\Sigma}_{s^*s}\boldsymbol{\Sigma}_{yy}^{-1}(\mathbf{y} - \mathbf{X}\hat{\boldsymbol{\beta}}_B).$$

All inferential tasks connected with the predicted signal $\mathbf{s}^*$ are solved by this distribution.

Example 1: Temperatures T_j^* shall be predicted as expected values at given times t_j^* with $j \in \{1, \ldots, r\}$ by means of the temperatures T_i measured at given times t_i with $i \in \{1, \ldots, n\}$. Let the errors of the measurements be independent and have variances which equal the variance factor σ^2. Let the prior information on σ^2 be given by

$$E(\sigma^2) = \sigma_p^2 \quad \text{and} \quad V(\sigma^2) = V_{\sigma^2}. \quad (335.8)$$

Let the systematic part or the trend be represented by a polynomial of degree m dependent on t_i and let with $\boldsymbol{\Sigma}_{ss} = (\sigma_{ik})$ the elements σ_{ik} of the covariance matrix $\boldsymbol{\Sigma}_{ss}$ of the signal be given by the covariance function $\sigma(t_i - t_k)$

$$\sigma_{ik} = \sigma(t_i - t_k) = ab^{(t_i - t_k)^2}, \quad (335.9)$$

where t_i and t_k refer to the times of the measurements and a and b are given constants. The covariance matrix $\boldsymbol{\Sigma}_{s^*s} = (\sigma_{j^*i})$ of the predicted signal and the signal and the covariance matrix $\boldsymbol{\Sigma}_{s^*s^*} = (\sigma_{j^*l^*})$ of the predicted signal are given by the same covariance function with the appropriate times t_j^*, t_l^* for the predicted signals and the time t_i for the measurements, thus

$$\sigma_{j^*i} = \sigma(t_j^* - t_i) = ab^{(t_j^* - t_i)^2} \quad (335.10)$$

and

$$\sigma_{j^*l^*} = \sigma(t_j^* - t_l^*) = ab^{(t_j^* - t_l^*)^2}. \quad (335.11)$$

In addition to the predicted temperature T_j^* the 95 percent confidence interval of the predicted signal contained besides the trend in T_j^* shall be determined.

The temperatures T_j^* are predicted as expected values by (335.5) with the Bayes estimates $\hat{\boldsymbol{\beta}}_B$ of the parameters $\boldsymbol{\beta}$ of the trend from (334.10)

$$\hat{\boldsymbol{\beta}}_B = \boldsymbol{\beta}_o = (\mathbf{X}'\boldsymbol{\Sigma}_{yy}^{-1}\mathbf{X})^{-1}\mathbf{X}'\boldsymbol{\Sigma}_{yy}^{-1}\mathbf{y},$$

where the matrix $\mathbf{X}$ is formed by the coefficients of the polynomials. With $\mathbf{X}=[\mathbf{x}_1,\ldots,\mathbf{x}_n]'$, where $\mathbf{x}_i'$ denotes a row of $\mathbf{X}$, we have

$$\mathbf{x}_i' = [1,t_i,t_i^2,\ldots,t_i^m] \quad \text{with} \quad i\in\{1,\ldots,n\}.$$

The matrix $\boldsymbol{\Sigma}_{yy}$ follows from (334.9) by

$$\boldsymbol{\Sigma}_{yy} = \boldsymbol{\Sigma}_{ss} + \boldsymbol{\Sigma}_{ee}$$

with $\boldsymbol{\Sigma}_{ss}$ from (335.9) and $\boldsymbol{\Sigma}_{ee}$ from

$$\boldsymbol{\Sigma}_{ee} = \mathbf{I},$$

since the errors are assumed as independent with equal variances σ^2. Finally we have

$$\mathbf{y} = (T_i) \quad \text{with} \quad i\in\{1,\ldots,n\}.$$

The Bayes estimate $\hat{\mathbf{s}}_B$ of the signal and the estimate $\hat{\mathbf{s}}_B^*$ of the predicted signal are obtained from (335.3) and (335.7) by

$$\hat{\mathbf{s}}_B = \mathbf{s}_o = \boldsymbol{\Sigma}_{ss}\boldsymbol{\Sigma}_{yy}^{-1}(\mathbf{y}-\mathbf{X}\hat{\boldsymbol{\beta}}_B)$$

and

$$\hat{\mathbf{s}}_B^* = \mathbf{s}_o^* = \boldsymbol{\Sigma}_{s^*s}\boldsymbol{\Sigma}_{yy}^{-1}(\mathbf{y}-\mathbf{X}\hat{\boldsymbol{\beta}}_B)$$

with $\boldsymbol{\Sigma}_{s^*s}$ from (335.10). The predicted temperatures $\hat{\mathbf{y}}_p=(\hat{T}_j^*)$ with $j\in\{1,\ldots,r\}$ follow from (335.5) by

$$\hat{\mathbf{y}}_p = \mathbf{X}^*\hat{\boldsymbol{\beta}}_B + \hat{\mathbf{s}}_B^*. \tag{335.12}$$

With $\mathbf{X}^*=[\mathbf{x}_1^*,\ldots,\mathbf{x}_r^*]'$, where $\mathbf{x}_j^{*\prime}$ denotes a row of $\mathbf{X}^*$, we have

$$\mathbf{x}_j^{*\prime} = [1,t_j^*,t_j^{*2},\ldots,t_j^{*m}] \quad \text{with} \quad j\in\{1,\ldots,r\}.$$

The confidence interval of the predicted signal s_j^* with $\mathbf{s}^*=(s_j^*)$ follows from the posterior distribution of s_j^*. It is obtained as the marginal distribution of (335.7). Thus, with (A22.10) and $\mathbf{s}_o^*=(s_{jo}^*)$ we obtain the posterior distribution

$$s_j^*|\mathbf{y} \sim t(s_{jo}^*,1/f,2p_o), \tag{335.13}$$

where

$$1/f = b_o(\boldsymbol{\Sigma}_{s^*s^*}-\boldsymbol{\Sigma}_{s^*s}\boldsymbol{\Sigma}_{yy}^{-1}\boldsymbol{\Sigma}_{ss^*}+(\boldsymbol{\Sigma}_{s^*s}-\boldsymbol{\Sigma}_{s^*s}\boldsymbol{\Sigma}_{yy}^{-1}\boldsymbol{\Sigma}_{ss})\boldsymbol{\Sigma}_{ee}^{-1}\mathbf{X}(\mathbf{X}'\boldsymbol{\Sigma}_{ee}^{-1}\mathbf{X}-\mathbf{X}'\boldsymbol{\Sigma}_{ee}^{-1}$$
$$(\boldsymbol{\Sigma}_{ss}-\boldsymbol{\Sigma}_{ss}\boldsymbol{\Sigma}_{yy}^{-1}\boldsymbol{\Sigma}_{ss})\boldsymbol{\Sigma}_{ee}^{-1}\mathbf{X})^{-1}\mathbf{X}'\boldsymbol{\Sigma}_{ee}^{-1}(\boldsymbol{\Sigma}_{ss^*}-\boldsymbol{\Sigma}_{ss}\boldsymbol{\Sigma}_{yy}^{-1}\boldsymbol{\Sigma}_{ss^*}))_{jj}/p_o$$

and from (335.3)

$$b_o = (2b+(\mathbf{y}-\mathbf{X}\boldsymbol{\beta}_o)'\Sigma_{yy}^{-1}\Sigma_{ss}\Sigma_{yy}^{-1}(\mathbf{y}-\mathbf{X}\boldsymbol{\beta}_o)+(\mathbf{y}-\mathbf{X}\boldsymbol{\beta}_o-\mathbf{s}_o)'\Sigma_{ee}^{-1}(\mathbf{y}-\mathbf{X}\boldsymbol{\beta}_o-\mathbf{s}_o))/2$$

$$p_o = (n+2p)/2, \quad p = (\sigma_p^2)^2/V_{\sigma^2} + 2, \quad b = (p-1)\sigma_p^2.$$

The distribution for s_j^* is the univariate t-distribution, whose density function is given in (A22.4). We apply the transformation (A22.5), to obtain a random variable having Student's t-distribution, so that tables of the t-distribution can be used. Hence, with $\hat{s}_{jB}^* = s_{jo}^*$ and (A22.6)

$$\sqrt{f}(s_j^*-\hat{s}_{jB}^*) \sim t(2p_o). \tag{335.14}$$

By introducing the quantity $t_{1-\alpha;2p_o}$, which was defined by (312.34) and which can be taken from a table of the t-distribution, we have

$$P(-t_{1-\alpha;2p_o} < \sqrt{f}(s_j^*-\hat{s}_{jB}^*) < t_{1-\alpha;2p_o}) = 1-\alpha$$

and obtain the confidence interval for s_j^* by

$$P(\hat{s}_{jB}^*-t_{1-\alpha;2p_o}/\sqrt{f} < s_j^* < \hat{s}_{jB}^*+t_{1-\alpha;2p_o}/\sqrt{f}) = 1-\alpha. \tag{335.15}$$

Finally the Bayes estimate $\hat{\sigma}_B^2$ of the variance factor σ^2 and its variance $V(\hat{\sigma}_B^2)$ follow from (333.13) by

$$\hat{\sigma}_B^2 = b_o/(p_o-1) \quad \text{and} \quad V(\hat{\sigma}_B^2) = b_o^2/((p_o-1)^2(p_o-2))$$

with b_o and p_o from (335.13). Δ

34 Linear Models with Unknown Variance and Covariance Components

341 Definition and Likelihood Function

We will now introduce the model which is equivalent to the Gauss-Markoff model with unknown variance and covariance components of the standard statistical techniques (Koch 1988a, p.264). In contrast to this model, where the unknown variance and covariance components are fixed quantities, the model for the Bayesian analysis introduces the unknown parameters as random quantities. Hence, the expected values of the observations and their covariance matrix have to be defined under the condition that the unknown parameters take on fixed values.

Definition: Let $\mathbf{X}$ be an n×u matrix of given coefficients, $\boldsymbol{\beta}$ a u×1 vector of unknown random parameters and $\mathbf{y}$ an n×1 random vector of observations. Let σ_i^2 and σ_{ij} be k unknown random parameters with $i \in \{1,\ldots,l\}$, $i<j\leq l$ and $1\leq k\leq l(l+1)/2$, which are called variance and covariance components, and let these k components be collected in the k×1 vector $\boldsymbol{\sigma}$. Let $E(\mathbf{y}|\boldsymbol{\beta})$ and $D(\mathbf{y}|\boldsymbol{\sigma})$ be the expected value and the covariance matrix of the observation vector $\mathbf{y}$ under the condition that the unknown parameter vectors $\boldsymbol{\beta}$ and $\boldsymbol{\sigma}$ are given. Then

$$E(\mathbf{y}|\boldsymbol{\beta}) = \mathbf{X}\boldsymbol{\beta} \quad \text{with} \quad D(\mathbf{y}|\boldsymbol{\sigma}) = \boldsymbol{\Sigma} = \sigma_1^2\mathbf{V}_1 + \sigma_{12}\mathbf{V}_2 + \ldots + \sigma_l^2\mathbf{V}_k$$

is called the *linear model with unknown variance and covariance components*, where the n×n covariance matrix $\boldsymbol{\Sigma}$ is positive definite and where the n×n matrices $\mathbf{V}_m$ with $m\in\{1,\ldots,k\}$ are known and symmetrical. (341.1)

The model with unknown variance and covariance components can be explained by the mixed model (331.1) with $\mathbf{Z}\boldsymbol{\gamma}=\mathbf{Z}_1\boldsymbol{\gamma}_1+\ldots+\mathbf{Z}_l\boldsymbol{\gamma}_l$. If the parameter vectors $\boldsymbol{\gamma}_i$ are unknown and unobservable and if their covariances $C(\boldsymbol{\gamma}_i,\boldsymbol{\gamma}_j)=\sigma_{ij}\mathbf{R}_{ij}$ are given except for the components σ_{ij}, then the covariance matrix for the observations $\mathbf{y}$ adopts the structure of (341.1) (Koch 1988a, p.264).

The mixed model is also used for a Bayesian analysis of variance components (Broemeling 1985, p.143). Such a model contains in addition to the variance components the parameter vectors $\boldsymbol{\gamma}_i$ as unknown parameters and the variance components enter the analysis by means of the prior distributions. The model (341.1) avoids the additional parameter vectors $\boldsymbol{\gamma}_i$, so that it is preferred here.

As before, we will assume the normal distribution for the observation vector $\mathbf{y}$ in (341.1) under the condition that the unknown parameter vectors $\boldsymbol{\beta}$ and $\boldsymbol{\sigma}$ are given

$$\mathbf{y}|\boldsymbol{\beta},\boldsymbol{\sigma} \sim N(\mathbf{X}\boldsymbol{\beta},\boldsymbol{\Sigma}), \tag{341.2}$$

where $\boldsymbol{\Sigma}$ according to (341.1) is a function of $\boldsymbol{\sigma}$. The likelihood function, that is the density function of the observations $\mathbf{y}$ given the parameters $\boldsymbol{\beta}$ and $\boldsymbol{\sigma}$, then follows from (A21.1) with

$$p(\mathbf{y}|\boldsymbol{\beta},\boldsymbol{\sigma}) = (2\pi)^{-n/2}(\det\boldsymbol{\Sigma})^{-1/2}\exp[-\frac{1}{2}(\mathbf{y}-\mathbf{X}\boldsymbol{\beta})'\boldsymbol{\Sigma}^{-1}(\mathbf{y}-\mathbf{X}\boldsymbol{\beta})]. \tag{341.3}$$

We are only interested in the parameter vector $\boldsymbol{\sigma}$. A noninformative prior is therefore introduced for the parameter vector $\boldsymbol{\beta}$, which for a linear model means a constant according to (312.2). Hence, the vector $\boldsymbol{\beta}$ may be directly integrated out of the density function (341.3). With the sufficient statistic $\hat{\boldsymbol{\beta}}$ for $\boldsymbol{\beta}$ from (224.4) in connection with (311.4)

$$\hat{\boldsymbol{\beta}} = (\mathbf{X}'\boldsymbol{\Sigma}^{-1}\mathbf{X})^{-1}\mathbf{X}'\boldsymbol{\Sigma}^{-1}\mathbf{y} \tag{341.4}$$

the exponent in (341.3) is transformed using the identity

$$\begin{aligned}(\mathbf{y}-\mathbf{X}\boldsymbol{\beta})'\boldsymbol{\Sigma}^{-1}(\mathbf{y}-\mathbf{X}\boldsymbol{\beta}) &= (\mathbf{y}-\mathbf{X}\hat{\boldsymbol{\beta}}-\mathbf{X}(\boldsymbol{\beta}-\hat{\boldsymbol{\beta}}))'\boldsymbol{\Sigma}^{-1}(\mathbf{y}-\mathbf{X}\hat{\boldsymbol{\beta}}-\mathbf{X}(\boldsymbol{\beta}-\hat{\boldsymbol{\beta}}))\\ &= (\mathbf{y}-\mathbf{X}\hat{\boldsymbol{\beta}})'\boldsymbol{\Sigma}^{-1}(\mathbf{y}-\mathbf{X}\hat{\boldsymbol{\beta}})+(\boldsymbol{\beta}-\hat{\boldsymbol{\beta}})'\mathbf{X}'\boldsymbol{\Sigma}^{-1}\mathbf{X}(\boldsymbol{\beta}-\hat{\boldsymbol{\beta}})\end{aligned}$$

because of

$$(\boldsymbol{\beta}-\hat{\boldsymbol{\beta}})'(\mathbf{X}'\boldsymbol{\Sigma}^{-1}\mathbf{y}-\mathbf{X}'\boldsymbol{\Sigma}^{-1}\mathbf{X}\hat{\boldsymbol{\beta}}) = 0.$$

The integration with respect to $\boldsymbol{\beta}$ gives because of (A21.2)

$$\int_{-\infty}^{\infty}\ldots\int_{-\infty}^{\infty}\exp[-\frac{1}{2}(\boldsymbol{\beta}-\hat{\boldsymbol{\beta}})'\mathbf{X}'\boldsymbol{\Sigma}^{-1}\mathbf{X}(\boldsymbol{\beta}-\hat{\boldsymbol{\beta}})]d\boldsymbol{\beta} = (2\pi)^{u/2}(\det\mathbf{X}'\boldsymbol{\Sigma}^{-1}\mathbf{X})^{-1/2}.$$

With

$$(\mathbf{y}-\mathbf{X}\hat{\boldsymbol{\beta}})'\boldsymbol{\Sigma}^{-1}(\mathbf{y}-\mathbf{X}\hat{\boldsymbol{\beta}}) = \mathbf{y}'\mathbf{W}\mathbf{y} \tag{341.5}$$

where

$$\mathbf{W} = \boldsymbol{\Sigma}^{-1}-\boldsymbol{\Sigma}^{-1}\mathbf{X}(\mathbf{X}'\boldsymbol{\Sigma}^{-1}\mathbf{X})^{-1}\mathbf{X}'\boldsymbol{\Sigma}^{-1} \tag{341.6}$$

we finally obtain after omitting the constants instead of (341.3) the likelihood function

$$p(\mathbf{y}|\boldsymbol{\sigma}) \propto (\det\boldsymbol{\Sigma}\ \det\mathbf{X}'\boldsymbol{\Sigma}^{-1}\mathbf{X})^{-1/2}\exp(-\mathbf{y}'\mathbf{W}\mathbf{y}/2), \tag{341.7}$$

which is only depending on the unknown parameter vector $\boldsymbol{\sigma}$.

The likelihood function (341.7) is very well suited for the Bayesian analysis of variance and covariance components out of the following reasons. By a linear transformation of the observation vector $\mathbf{y}$ we may factor the likelihood function (341.3) into two products

L_1 and L_2, where L_1 is identical with the likelihood function (341.7) and therefore is only dependent on the parameter vector $\boldsymbol{\sigma}$ (Koch 1987). L_2 depends on $\boldsymbol{\beta}$ and $\boldsymbol{\sigma}$ and leads in a maximum likelihood estimate to the well-known estimator (341.4), if $\boldsymbol{\sigma}$ is assumed as known. Using (341.7) in a maximum likelihood estimate gives the following estimator of $\boldsymbol{\sigma}$ (Koch 1986). Instead of $\boldsymbol{\sigma}$ we estimate $\bar{\boldsymbol{\sigma}}$ by $\hat{\bar{\boldsymbol{\sigma}}}$ with

$$\hat{\bar{\boldsymbol{\sigma}}} = \bar{\mathbf{S}}^{-1}\bar{\mathbf{q}}, \qquad (341.8)$$

where

$$\bar{\boldsymbol{\sigma}} = [\ldots,\bar{\sigma}_i^2,\ldots,\bar{\sigma}_{ij},\ldots]', \quad \boldsymbol{\sigma} = [\ldots,\alpha_i^2\bar{\sigma}_i^2,\ldots,\alpha_{ij}\bar{\sigma}_{ij},\ldots]'$$

$$\bar{\mathbf{S}} = (\mathrm{tr}\bar{\mathbf{W}}\bar{\mathbf{V}}_i\bar{\mathbf{W}}\bar{\mathbf{V}}_j) \quad \text{for} \quad i,j\in\{1,\ldots,k\}, \quad \bar{\mathbf{q}} = (\mathbf{y}'\bar{\mathbf{W}}\bar{\mathbf{V}}_i\bar{\mathbf{W}}\mathbf{y}) \quad \text{for} \quad i\in\{1,\ldots,k\}$$

$$\mathbf{V}_m = \bar{\mathbf{V}}_m/\alpha_i^2 \quad \text{or} \quad \mathbf{V}_m = \bar{\mathbf{V}}_m/\alpha_{ij}, \quad \bar{\mathbf{W}} = \boldsymbol{\Sigma}_o^{-1} - \boldsymbol{\Sigma}_o^{-1}\mathbf{X}(\mathbf{X}'\boldsymbol{\Sigma}_o^{-1}\mathbf{X})^{-1}\mathbf{X}'\boldsymbol{\Sigma}_o^{-1}$$

$$\boldsymbol{\Sigma}_o = \sum_{m=1}^{k}\bar{\mathbf{V}}_m.$$

Thus, for the estimation, approximate values α_i^2 and α_{ij} are split off the variance components σ_i^2 and σ_{ij} to obtain $\bar{\sigma}_i^2$ and $\bar{\sigma}_{ij}$, which can be assumed as having values close to one. The approximate values are absorbed in the matrices $\mathbf{V}_m$, which then become $\bar{\mathbf{V}}_m$. The estimator (341.8) is not only a maximum likelihood estimator, but also a best invariant quadratic unbiased estimator (Koch 1988a, p.270). It is generally applied when estimating variance and covariance components for geodetic applications. Hence, when introducing the likelihood function (341.7) for a Bayesian analysis, the estimate (232.4) will give approximately the same results as (341.8), if the prior information is noninformative or is available with large variances only.

342 Noninformative Priors

We will assume a noninformative prior density function for the unknown parameter vector $\boldsymbol{\sigma}$ and apply Jeffrey's invariance principle to derive it. Thus, according to (222.11) we take the natural logarithm of the likelihood function (341.7) and obtain after omitting the constant

$$\ln p(\mathbf{y}|\boldsymbol{\sigma}) \propto -\ln\det\boldsymbol{\Sigma} - \ln\det\mathbf{X}'\boldsymbol{\Sigma}^{-1}\mathbf{X} - \mathbf{y}'\mathbf{W}\mathbf{y}. \qquad (342.1)$$

For the differentiation of (342.1) with respect to the components of $\boldsymbol{\sigma}$ we apply the following two rules (see, for instance, Kubik 1970). Let $\mathbf{A}$ be a regular symmetric m×m matrix, which is a function of the vector $\mathbf{a}$ with $\mathbf{a}=(a_i)$, then

$$\partial \ln\det\mathbf{A}/\partial a_i = \mathrm{tr}\mathbf{A}^{-1}\partial\mathbf{A}/\partial a_i \tag{342.2}$$

and

$$\partial\mathbf{A}^{-1}/\partial a_i = -\mathbf{A}^{-1}\partial\mathbf{A}/\partial a_i\mathbf{A}^{-1}. \tag{342.3}$$

To prove (342.2) we use $\mathbf{A}=\mathbf{Y}\boldsymbol{\Lambda}\mathbf{Y}'$ with $\mathbf{Y}'\mathbf{Y}=\mathbf{I}$, where $\boldsymbol{\Lambda}$ is the diagonal matrix of the eigenvalues λ_i of $\mathbf{A}$ and $\mathbf{Y}$ the matrix of the eigenvectors (Koch 1988a, p.53). We find

$$\ln\det\mathbf{A} = \ln\det\boldsymbol{\Lambda} = \ln\prod_{j=1}^{m}\lambda_j = \sum_{j=1}^{m}\ln\lambda_j$$

and

$$\partial\ln\det\mathbf{A}/\partial a_i = \sum_{j=1}^{m}\lambda_j^{-1}\partial\lambda_j/\partial a_i = \mathrm{tr}\boldsymbol{\Lambda}^{-1}\partial\boldsymbol{\Lambda}/\partial a_i.$$

In addition, we obtain from $\mathbf{A}=\mathbf{Y}\boldsymbol{\Lambda}\mathbf{Y}'$

$$\partial\mathbf{A}/\partial a_i = \partial\mathbf{Y}/\partial a_i\boldsymbol{\Lambda}\mathbf{Y}' + \mathbf{Y}\partial\boldsymbol{\Lambda}/\partial a_i\mathbf{Y}' + \mathbf{Y}\boldsymbol{\Lambda}\partial\mathbf{Y}'/\partial a_i$$

and from $\mathbf{A}^{-1}=\mathbf{Y}\boldsymbol{\Lambda}^{-1}\mathbf{Y}'$

$$\mathrm{tr}\mathbf{A}^{-1}\partial\mathbf{A}/\partial a_i = \mathrm{tr}\mathbf{Y}'\partial\mathbf{Y}/\partial a_i + \mathrm{tr}\boldsymbol{\Lambda}^{-1}\partial\boldsymbol{\Lambda}/\partial a_i + \mathrm{tr}\mathbf{Y}\partial\mathbf{Y}'/\partial a_i.$$

Finally $\partial\mathbf{Y}'/\partial a_i\mathbf{Y}+\mathbf{Y}'\partial\mathbf{Y}/\partial a_i=\mathbf{0}$ because of $\mathbf{Y}'\mathbf{Y}=\mathbf{I}$, so that we find

$$\mathrm{tr}\mathbf{A}^{-1}\partial\mathbf{A}/\partial a_i = \mathrm{tr}\boldsymbol{\Lambda}^{-1}\partial\boldsymbol{\Lambda}/\partial a_i,$$

which gives (342.2).

From $\mathbf{A}\mathbf{A}^{-1}=\mathbf{I}$ we obtain

$$\partial\mathbf{A}/\partial a_i\mathbf{A}^{-1} + \mathbf{A}\partial\mathbf{A}^{-1}/\partial a_i = \mathbf{0}$$

and

$$\partial\mathbf{A}^{-1}/\partial a_i = -\mathbf{A}^{-1}\partial\mathbf{A}/\partial a_i\mathbf{A}^{-1},$$

which is identical with (342.3).

By applying (342.2) and (342.3) to the derivative of (342.1) we find with $\boldsymbol{\sigma}=(\sigma_i)$

$$\begin{aligned}\partial\ln p(\mathbf{y}|\boldsymbol{\sigma})/\partial\sigma_i \propto & -\mathrm{tr}\boldsymbol{\Sigma}^{-1}\partial\boldsymbol{\Sigma}/\partial\sigma_i + \mathrm{tr}(\mathbf{X}'\boldsymbol{\Sigma}^{-1}\mathbf{X})^{-1}\mathbf{X}'\boldsymbol{\Sigma}^{-1}\partial\boldsymbol{\Sigma}/\partial\sigma_i\boldsymbol{\Sigma}^{-1}\mathbf{X}\\ & -\mathbf{y}'[-\boldsymbol{\Sigma}^{-1}\partial\boldsymbol{\Sigma}/\partial\sigma_i\boldsymbol{\Sigma}^{-1}+\boldsymbol{\Sigma}^{-1}\partial\boldsymbol{\Sigma}/\partial\sigma_i\boldsymbol{\Sigma}^{-1}\mathbf{X}(\mathbf{X}'\boldsymbol{\Sigma}^{-1}\mathbf{X})^{-1}\mathbf{X}'\boldsymbol{\Sigma}^{-1}\\ & -\boldsymbol{\Sigma}^{-1}\mathbf{X}(\mathbf{X}'\boldsymbol{\Sigma}^{-1}\mathbf{X})^{-1}\mathbf{X}'\boldsymbol{\Sigma}^{-1}\partial\boldsymbol{\Sigma}/\partial\sigma_i\boldsymbol{\Sigma}^{-1}\mathbf{X}(\mathbf{X}'\boldsymbol{\Sigma}^{-1}\mathbf{X})^{-1}\mathbf{X}'\boldsymbol{\Sigma}^{-1}\\ & +\boldsymbol{\Sigma}^{-1}\mathbf{X}(\mathbf{X}'\boldsymbol{\Sigma}^{-1}\mathbf{X})^{-1}\mathbf{X}'\boldsymbol{\Sigma}^{-1}\partial\boldsymbol{\Sigma}/\partial\sigma_i\boldsymbol{\Sigma}^{-1}]\mathbf{y} = -\mathrm{tr}\mathbf{W}\partial\boldsymbol{\Sigma}/\partial\sigma_i + \mathbf{y}'\mathbf{W}\partial\boldsymbol{\Sigma}/\partial\sigma_i\mathbf{W}\mathbf{y},\end{aligned}$$

where

$$\partial\boldsymbol{\Sigma}/\partial\sigma_i = \mathbf{V}_i \quad \text{with} \quad i\in\{1,\ldots,k\}$$

because of (341.1). Thus,

$$\partial \ln p(\mathbf{y}|\boldsymbol{\sigma})/\partial\sigma_i \propto -\operatorname{tr}\mathbf{W}\mathbf{V}_i + \mathbf{y}'\mathbf{W}\mathbf{V}_i\mathbf{W}\mathbf{y}$$

and

$$\partial\mathbf{W}/\partial\sigma_i = -\mathbf{W}\mathbf{V}_i\mathbf{W}.$$

The second derivative follows with

$$\partial^2 \ln p(\mathbf{y}|\boldsymbol{\sigma})/\partial\sigma_i\partial\sigma_j \propto \operatorname{tr}\mathbf{W}\mathbf{V}_j\mathbf{W}\mathbf{V}_i - 2\mathbf{y}'\mathbf{W}\mathbf{V}_j\mathbf{W}\mathbf{V}_i\mathbf{W}\mathbf{y}.$$

Now we have to take the expected value of this derivative and obtain with (341.1)

$$E(\mathbf{y}\mathbf{y}'|\boldsymbol{\beta},\boldsymbol{\sigma}) = \boldsymbol{\Sigma} + \mathbf{X}\boldsymbol{\beta}\boldsymbol{\beta}'\mathbf{X}'$$

(Koch 1988a, p.114) and therefore

$$E(\mathbf{y}'\mathbf{W}\mathbf{V}_j\mathbf{W}\mathbf{V}_i\mathbf{W}\mathbf{y}) = \operatorname{tr}\mathbf{W}\mathbf{V}_j\mathbf{W}\mathbf{V}_i\mathbf{W}E(\mathbf{y}\mathbf{y}') = \operatorname{tr}\mathbf{W}\mathbf{V}_j\mathbf{W}\mathbf{V}_i$$

because of

$$\mathbf{W} = \mathbf{W}\boldsymbol{\Sigma}\mathbf{W} \quad \text{and} \quad \mathbf{W}\mathbf{X} = \mathbf{0}.$$

Thus, we finally obtain with (222.10) the noninformative density function $p(\boldsymbol{\sigma})$ of the parameter vector $\boldsymbol{\sigma}$ by

$$p(\boldsymbol{\sigma}) \propto (\det\mathbf{S})^{1/2}, \tag{342.4}$$

where

$$\mathbf{S} = (\operatorname{tr}\mathbf{W}\mathbf{V}_i\mathbf{W}\mathbf{V}_j) \quad \text{for} \quad i,j\in\{1,\ldots,k\}.$$

With the prior density (342.4) and the likelihood function (341.7) the posterior density $p(\boldsymbol{\sigma}|\mathbf{y})$ of the variance and covariance components σ_i^2 and σ_{ij} contained in $\boldsymbol{\sigma}$ follows with Bayes' theorem (211.1) from

$$p(\boldsymbol{\sigma}|\mathbf{y}) \propto (\det\mathbf{S})^{1/2}(\det\boldsymbol{\Sigma}\,\det\mathbf{X}'\boldsymbol{\Sigma}^{-1}\mathbf{X})^{-1/2}\exp(-\mathbf{y}'\mathbf{W}\mathbf{y}/2). \tag{342.5}$$

This distribution depends through the matrix $\boldsymbol{\Sigma}$ and the matrices $\mathbf{S}$ and $\mathbf{W}$ on the unknown variance and covariance components σ_i^2 and σ_{ij} in $\boldsymbol{\sigma}$. By integrating over the domain of $\boldsymbol{\sigma}$, the normalization factor is determined according to (211.5). Unfortunately, this integration could not be solved analytically. The same holds true for the integration (231.5), which gives the Bayes estimate $\hat{\boldsymbol{\sigma}}_B$ for the variance and covariance components in $\boldsymbol{\sigma}$, and the integration leading to the confidence region (241.2) or the hypothesis test for the variance and covariance components. Hence, the numerical methods of Section 27 have to be applied, to compute the Bayes estimates for the variance and covariance components, their confidence regions, and to test hypotheses.

Example 1: The length s of a straight line has been measured by two instruments such, that two independent sets of observations are obtained. The observations of the first set are collected in the $n_1 \times 1$ vector $\mathbf{y}_1$ and of the second set in the $n_2 \times 1$ vector $\mathbf{y}_2$. The observations in $\mathbf{y}_1$ are also independent and have equal weights p_1, the observations in $\mathbf{y}_2$ are independent, too, and have equal weights p_2. The covariance matrix of $\mathbf{y}=[\mathbf{y}_1', \mathbf{y}_2']'$, which is a diagonal matrix, shall be expressed by the variance component σ_1^2 times the inverse $1/p_1$ of the weights for $\mathbf{y}_1$ and by σ_2^2 times $1/p_2$ for $\mathbf{y}_2$. The variance components σ_1^2 and σ_2^2 are the unknown parameters, whose Bayes estimates and whose 95 per cent confidence intervals have to be computed.

From (341.1) we obtain the following model for the example

$$E(\mathbf{y}|\boldsymbol{\beta}) = \mathbf{X}\boldsymbol{\beta} \quad \text{with} \quad D(\mathbf{y}|\boldsymbol{\sigma}) = \boldsymbol{\Sigma} = \sigma_1^2\mathbf{V}_1 + \sigma_2^2\mathbf{V}_2, \tag{342.6}$$

where

$$\mathbf{y} = \begin{bmatrix} \mathbf{y}_1 \\ \mathbf{y}_2 \end{bmatrix}, \quad \boldsymbol{\beta} = s$$

$$\mathbf{X} = \begin{bmatrix} \mathbf{e}_1 \\ \mathbf{e}_2 \end{bmatrix}, \quad \mathbf{e}_1 = [1,1,\ldots,1]', \quad \mathbf{e}_2 = [1,1,\ldots,1]'$$

$$\mathbf{V}_1 = (1/p_1)\begin{bmatrix} \mathbf{I}_{n1} & \mathbf{0} \\ \mathbf{0} & \mathbf{0} \end{bmatrix}, \quad \mathbf{V}_2 = (1/p_2)\begin{bmatrix} \mathbf{0} & \mathbf{0} \\ \mathbf{0} & \mathbf{I}_{n2} \end{bmatrix}.$$

The vectors $\mathbf{e}_1$ and $\mathbf{e}_2$ are $n_1 \times 1$ and $n_2 \times 1$ vectors and $\mathbf{I}_{n1}$ and $\mathbf{I}_{n2}$ are $n_1 \times n_1$ and $n_2 \times n_2$ identity matrices, respectively.

The posterior density function (342.5) will be applied for the statistical inference on the variance components σ_1^2 and σ_2^2 of the model (342.6). To compute the density values for σ_1^2 and σ_2^2, we need the following matrices

$$\boldsymbol{\Sigma} = (\sigma_1^2/p_1)\begin{bmatrix} \mathbf{I}_{n1} & \mathbf{0} \\ \mathbf{0} & \mathbf{0} \end{bmatrix} + (\sigma_2^2/p_2)\begin{bmatrix} \mathbf{0} & \mathbf{0} \\ \mathbf{0} & \mathbf{I}_{n2} \end{bmatrix}$$

$$\boldsymbol{\Sigma}^{-1} = \begin{bmatrix} (p_1/\sigma_1^2)\mathbf{I}_{n1} & \mathbf{0} \\ \mathbf{0} & (p_2/\sigma_2^2)\mathbf{I}_{n2} \end{bmatrix}, \quad \boldsymbol{\Sigma}^{-1}\mathbf{X} = \begin{bmatrix} (p_1/\sigma_1^2)\ \mathbf{e}_1 \\ (p_2/\sigma_2^2)\ \mathbf{e}_2 \end{bmatrix}$$

$$\mathbf{X}'\boldsymbol{\Sigma}^{-1}\mathbf{X} = n_1p_1/\sigma_1^2 + n_2p_2/\sigma_2^2$$

$$\mathbf{\Sigma}^{-1}\mathbf{X}(\mathbf{X}'\mathbf{\Sigma}^{-1}\mathbf{X})^{-1}\mathbf{X}'\mathbf{\Sigma}^{-1} = (n_1p_1/\sigma_1^2+n_2p_2/\sigma_2^2)^{-1}$$

$$\begin{bmatrix} (p_1/\sigma_1^2)^2\mathbf{e}_1\mathbf{e}_1' & (p_1p_2/\sigma_1^2\sigma_2^2)\mathbf{e}_1\mathbf{e}_2' \\ (p_1p_2/\sigma_1^2\sigma_2^2)\mathbf{e}_2\mathbf{e}_1' & (p_2/\sigma_2^2)^2\mathbf{e}_2\mathbf{e}_2' \end{bmatrix}$$

$$\mathbf{W} = \begin{bmatrix} \mathbf{W}_{11} & \mathbf{W}_{12} \\ \mathbf{W}_{21} & \mathbf{W}_{22} \end{bmatrix}, \quad \mathbf{W}_{21} = \mathbf{W}_{12}'$$

$$\mathbf{W}_{11} = (p_1/\sigma_1^2)\mathbf{I}_{n1} - (p_1/\sigma_1^2)^2\mathbf{e}_1\mathbf{e}_1'/(n_1p_1/\sigma_1^2+n_2p_2/\sigma_2^2)$$

$$\mathbf{W}_{12} = -(p_1p_2/\sigma_1^2\sigma_2^2)\mathbf{e}_1\mathbf{e}_2'/(n_1p_1/\sigma_1^2+n_2p_2/\sigma_2^2)$$

$$\mathbf{W}_{22} = (p_2/\sigma_2^2)\mathbf{I}_{n2} - (p_2/\sigma_2^2)^2\mathbf{e}_2\mathbf{e}_2'/(n_1p_1/\sigma_1^2+n_2p_2/\sigma_2^2)$$

$$\mathbf{W}\mathbf{V}_1 = \begin{bmatrix} \mathbf{W}_{11}/p_1 & \mathbf{0} \\ \mathbf{W}_{21}/p_1 & \mathbf{0} \end{bmatrix}, \quad \mathbf{W}\mathbf{V}_2 = \begin{bmatrix} \mathbf{0} & \mathbf{W}_{12}/p_2 \\ \mathbf{0} & \mathbf{W}_{22}/p_2 \end{bmatrix}.$$

The elements of the matrix **S** follow from (342.4) with

$$\mathbf{S} = (s_{ij}) = (\mathrm{tr}\mathbf{W}\mathbf{V}_i\mathbf{W}\mathbf{V}_j) \quad \text{for} \quad i,j\in\{1,2\},$$

where

$$s_{11} = n_1/(\sigma_1^2)^2 - 2n_1p_1/(\sigma_1^2)^3(n_1p_1/\sigma_1^2+n_2p_2/\sigma_2^2)$$
$$+ [n_1p_1/(\sigma_1^2)^2(n_1p_1/\sigma_1^2+n_2p_2/\sigma_2^2)]^2$$

$$s_{12} = s_{21} = n_1n_2p_1p_2/[\sigma_1^2\sigma_2^2(n_1p_1/\sigma_1^2+n_2p_2/\sigma_2^2)]^2$$

$$s_{22} = n_2/(\sigma_2^2)^2 - 2n_2p_2/(\sigma_2^2)^3(n_1p_1/\sigma_1^2+n_2p_2/\sigma_2^2)$$
$$+ [n_2p_2/(\sigma_2^2)^2(n_1p_1/\sigma_1^2+n_2p_2/\sigma_2^2)]^2.$$

The determinants of **S**, $\mathbf{\Sigma}$ and $\mathbf{X}'\mathbf{\Sigma}^{-1}\mathbf{X}$ are computed by

$$\det\mathbf{S} = s_{11}s_{22} - s_{12}^2$$
$$\det\mathbf{\Sigma} = (\sigma_1^2/p_1)^{n_1}(\sigma_2^2/p_2)^{n_2}$$
$$\det\mathbf{X}'\mathbf{\Sigma}^{-1}\mathbf{X} = n_1p_1/\sigma_1^2 + n_2p_2/\sigma_2^2. \tag{342.7}$$

The quadratic form $\mathbf{y}'\mathbf{W}\mathbf{y}$ follows with (341.5) by means of the vector $\hat{\mathbf{e}}=\mathbf{X}\hat{\boldsymbol{\beta}}-\mathbf{y}$ of the residuals. We obtain with (341.4), $\mathbf{y}_1=(y_{i1})$ and $\mathbf{y}_2=(y_{i2})$ the estimate $\hat{s}$ of the length s by

$$\hat{s} = (n_1p_1/\sigma_1^2+n_2p_2/\sigma_2^2)^{-1}[(p_1/\sigma_1^2)\sum_{i=1}^{n_1} y_{i1} + (p_2/\sigma_2^2)\sum_{i=1}^{n_2} y_{i2}]$$

and the vector $\hat{\mathbf{e}}$ of the residuals by

$$\hat{\mathbf{e}} = \begin{bmatrix}\mathbf{e}_1\\ \mathbf{e}_2\end{bmatrix}\hat{s} - \begin{bmatrix}\mathbf{y}_1\\ \mathbf{y}_2\end{bmatrix}.$$

The quadratic form of the residuals therefore follows with

$$\mathbf{y}'\mathbf{W}\mathbf{y} = (p_1/\sigma_1^2)\sum_{i=1}^{n_1}(\hat{s}-y_{i1})^2 + (p_2/\sigma_2^2)\sum_{i=1}^{n_2}(\hat{s}-y_{i2})^2. \tag{342.8}$$

With the determinants (342.7) and the quadratic form (342.8) of the residuals all quantities are available, to compute the density values of the posterior distribution for σ_1^2 and σ_2^2 from (342.5).

In addition, the variance components shall be estimated by (341.8). We have

$$\bar{\mathbf{V}}_1 = (\alpha_1^2/p_1)\begin{bmatrix}\mathbf{I}_{n1} & \mathbf{0}\\ \mathbf{0} & \mathbf{0}\end{bmatrix},\quad \bar{\mathbf{V}}_2 = (\alpha_2^2/p_2)\begin{bmatrix}\mathbf{0} & \mathbf{0}\\ \mathbf{0} & \mathbf{I}_{n2}\end{bmatrix}$$

and

$$\mathbf{\Sigma}_o = (\alpha_1^2/p_1)\begin{bmatrix}\mathbf{I}_{n1} & \mathbf{0}\\ \mathbf{0} & \mathbf{0}\end{bmatrix} + (\alpha_2^2/p_2)\begin{bmatrix}\mathbf{0} & \mathbf{0}\\ \mathbf{0} & \mathbf{I}_{n2}\end{bmatrix}.$$

The elements of the matrix $\bar{\mathbf{W}}$ therefore follow from the elements of the matrix $\mathbf{W}$ by replacing σ_1^2 and σ_2^2 with α_1^2 and α_2^2. The same applies for the estimate $\hat{s}$. Furthermore, we have

$$\bar{\mathbf{W}}\bar{\mathbf{V}}_1 = \begin{bmatrix}\alpha_1^2\bar{\mathbf{W}}_{11}/p_1 & \mathbf{0}\\ \alpha_1^2\bar{\mathbf{W}}_{21}/p_1 & \mathbf{0}\end{bmatrix},\quad \bar{\mathbf{W}}\bar{\mathbf{V}}_2 = \begin{bmatrix}\mathbf{0} & \alpha_2^2\bar{\mathbf{W}}_{12}/p_2\\ \mathbf{0} & \alpha_2^2\bar{\mathbf{W}}_{22}/p_2\end{bmatrix}.$$

The elements of the matrix $\bar{\mathbf{S}}$ are obtained with $\bar{\mathbf{S}}=(\bar{s}_{ij})$ by

$$\bar{s}_{11} = n_1 - 2n_1p_1/\alpha_1^2(n_1p_1/\alpha_1^2+n_2p_2/\alpha_2^2) + [n_1p_1/\alpha_1^2(n_1p_1/\alpha_1^2+n_2p_2/\alpha_2^2)]^2$$

$$\bar{s}_{12} = \bar{s}_{21} = n_1n_2p_1p_2/\alpha_1^2\alpha_2^2(n_1p_1/\alpha_1^2+n_2p_2/\alpha_2^2)^2$$

$$\bar{s}_{22} = n_2 - 2n_2p_2/\alpha_2^2(n_1p_1/\alpha_1^2+n_2p_2/\alpha_2^2) + [n_2p_2/\alpha_2^2(n_1p_1/\alpha_1^2+n_2p_2/\alpha_2^2)]^2. \tag{342.9}$$

The vector $\bar{\mathbf{q}}$ in (341.8) is computed from (Koch 1988a, p.269)

$$\bar{\mathbf{q}} = (\bar{q}_i) = (\hat{\mathbf{e}}'\mathbf{\Sigma}_o^{-1}\bar{\mathbf{V}}_i\mathbf{\Sigma}_o^{-1}\hat{\mathbf{e}}) \quad \text{for} \quad i\in\{1,2\}$$

with

$$\bar{q}_1 = (p_1/\alpha_1^2) \sum_{i=1}^{n_1} (\hat{s}-y_{i1})^2, \quad \bar{q}_2 = (p_2/\alpha_2^2) \sum_{i=1}^{n_2} (\hat{s}-y_{i2})^2. \tag{342.10}$$

With the elements of $\bar{\mathbf{S}}$ and $\bar{\mathbf{q}}$ the estimates of the variance components are obtained from (341.8).

The following measurements in units of millimeters have been taken

$$\mathbf{y}_1 = \begin{bmatrix} 8\ 264. \\ 8\ 266. \\ 8\ 267. \\ 8\ 264. \\ 8\ 268. \\ 8\ 263. \\ 8\ 265. \end{bmatrix} \quad \mathbf{y}_2 = \begin{bmatrix} 8\ 266. \\ 8\ 262. \\ 8\ 264. \\ 8\ 267. \\ 8\ 269. \end{bmatrix}.$$

The weights p_1 and p_2 of the two sets of measurements are given in units of 1/millimeter^2 by

$$p_1 = 0.250, \quad p_2 = 0.111.$$

By starting with

$$\alpha_1^2 = 0.8, \quad \alpha_2^2 = 0.8$$

the estimation (341.8) together with (342.9) and (342.10) has been iteratively applied with choosing $\hat{\sigma}_1^2=\alpha_1^2\hat{\bar{\sigma}}_1^2$ and $\hat{\sigma}_2^2=\alpha_2^2\hat{\bar{\sigma}}_2^2$ as new approximate values. After five iterations we obtain $\hat{\bar{\sigma}}_1^2=1.0000$ and $\hat{\bar{\sigma}}_2^2=1.0000$ and the iterated estimates of σ_1^2 and σ_2^2

$$\hat{\sigma}_1^2 = 0.777, \quad \hat{\sigma}_2^2 = 0.691 \tag{342.11}$$

with the standard deviations of the estimates $\hat{\sigma}_1^2$ and $\hat{\sigma}_2^2$ (Koch 1988a, p.273)

$$(\hat{V}(\hat{\sigma}_1^2))^{1/2} = 0.45, \quad (\hat{V}(\hat{\sigma}_2^2))^{1/2} = 0.46. \tag{342.12}$$

The posterior density function (342.5) together with (342.7) and (342.8) was then used to compute the Bayes estimate of the variance component σ_1^2 and to obtain its confidence region. First the marginal posterior density function for σ_1^2 was computed from (342.5) by applying (273.2) with i=500 and j=500. Thus, for each of the 500 random numbers with uniform distributions generated for σ_1^2, 500 random numbers with uniform distributions were generated for σ_2^2, whose density values were summed according to (273.2).

The marginal posterior density values for σ_1^2 were then used to compute the Bayes estimate $\hat{\sigma}_{1B}^2$ of σ_1^2 from (272.3) and the confidence interval for σ_1^2 from (272.6). Since the marginal posterior density function is univariate, the generated values for σ_1^2 were ordered according to increasing values and along with them the density values. Then ten density values between all consecutive pairs of density values were added by linear interpolation and the confidence limits were computed from (272.6) together with (272.7) by starting from the density values of the minimum and maximum values for σ_1^2 and by numerically checking whether the inequality in (241.2) was fulfilled.

The interval in (271.4) for the coordinate axis of σ_1^2, which defines the domain of the integration, was set up such that the interval of the integration is ten to twenty per cent wider than the confidence interval of σ_1^2. This ensures that in analogy to (272.1) only the density values are neglected, which cease to contribute to the integration. By computing the confidence interval for σ_2^2 as described below, the limits of the integration for σ_2^2 were determined.

For each of the two following intervals defining the domain of the integration

$$0.05 < \sigma_1^2 < 2.4, \quad 0.05 < \sigma_2^2 < 2.4$$

and

$$0.05 < \sigma_1^2 < 2.2, \quad 0.05 < \sigma_2^2 < 2.2 \qquad (342.13)$$

the two sets of 500 times 500 random numbers were generated and $\hat{\sigma}_{1B}^2$ and the confidence limits for σ_1^2 were computed. The results are given in the table (342.14). The confidence interval for σ_1^2 is not symmetrical with respect to the Bayes estimate $\hat{\sigma}_{1B}^2$, since the marginal posterior density function for σ_1^2 is not symmetrical. Its maximum lies in the vicinity of the iterated estimate (342.11), which results from a maximum likelihood estimate based on the likelihood function (341.7). The interval [0.33, 1.23] defined by

	$\hat{\sigma}_{1B}^2$	Confidence Interval
	0.94	$0.25 < \sigma_1^2 < 1.91$
	0.94	$0.25 < \sigma_1^2 < 1.95$
	0.92	$0.26 < \sigma_1^2 < 1.83$
	0.91	$0.26 < \sigma_1^2 < 1.88$
Mean	0.93	$0.26 < \sigma_1^2 < 1.89$

(342.14)

the estimate (342.11) of σ_1^2 and its standard deviation (342.12) lies within the confidence interval.

The same procedure as described for the variance component σ_1^2 was then applied to compute the Bayes estimate and the confidence interval for the variance component σ_2^2. The results for the two intervals (342.13) are presented in the table (342.15). Due to the different weights for the observation vectors $\mathbf{y}_1$ and $\mathbf{y}_2$ and due to the fact that the obser-

	$\hat{\sigma}^2_{2B}$	Confidence Interval
	0.86	$0.19 < \sigma_2^2 < 1.97$
	0.87	$0.20 < \sigma_2^2 < 1.87$
	0.85	$0.20 < \sigma_2^2 < 1.85$
	0.86	$0.20 < \sigma_2^2 < 1.84$
Mean	0.86	$0.20 < \sigma_2^2 < 1.88$

(342.15)

vation vector $\mathbf{y}_2$, for which σ_2^2 was introduced, contains only two observations less than $\mathbf{y}_1$, the results of the table (342.15) are very similar to those of (342.14). Δ

343 Informative Priors

We will now assume informative priors for the unknown variance and covariance components σ_i^2 and σ_{ij}. The prior density functions shall be determined by the given expected values and variances of the k variance and covariance components, that is by

$$E(\sigma_i^2) = \mu_{\sigma i} \quad \text{and} \quad V(\sigma_i^2) = V_{\sigma i} \quad \text{for} \quad i \in \{1,\dots,l\}$$
$$E(\sigma_{ij}) = \mu_{\sigma ij} \quad \text{and} \quad V(\sigma_{ij}) = V_{\sigma ij} \quad \text{for} \quad i < j \le l, \quad l \le k \le l(l+1)/2. \tag{343.1}$$

Like the variance factor σ^2 of the linear model (311.1), the variance components σ_i^2 take on positive values only. In the linear model the gamma distribution (A12.1) via the normal-gamma distribution (A23.1) is used as prior for the weight parameter $\tau=1/\sigma^2$. This corresponds to the inverted gamma distribution (A13.1) for σ^2. Thus, the gamma distribution would be the appropriate prior distribution for $1/\sigma_i^2$. This distribution has been used as prior for the Bayesian analysis of variance components in the mixed model (Broemeling 1985, p.146; Bunke and Bunke 1986, p.473), while the inverted gamma distribution (A13.1) of the variance components σ_i^2 was applied for the Bayesian analysis in the model (341.1) (Koch 1988b). By assuming the l variance components σ_i^2 as being independent, the prior $p(\sigma_i^2)_l$ for the l components σ_i^2 is obtained from the inverted gamma distribution (A13.1) with neglecting the constants by

$$p(\sigma_i^2)_l \propto \prod_{i=1}^{l} (1/\sigma_i^2)^{p_i+1} \exp(-b_i/\sigma_i^2). \tag{343.2}$$

The product is taken over the l variance components σ_i^2, and b_i and p_i are the parameters of the distributions for the individual variance components. They follow with (343.1) from (A13.2) with

$$p_i = \mu^2_{\sigma i}/V_{\sigma i} + 2, \quad b_i = (p_i - 1)\mu_{\sigma i}. \tag{343.3}$$

Provided $V_{\sigma i} < \mu^2_{\sigma i}$ holds, then a very meaningful choice for the prior distribution of the variance components σ^2_i is the truncated normal distribution (223.8), which is based on the maximum entropy principle. By again assuming the l variance components σ^2_i as being independent, we obtain the prior $p(\sigma^2_i)_l$ for the l components σ^2_i from (223.8) with neglecting the constants by

$$p(\sigma^2_i)_l \propto \prod_{i=1}^{l} \exp(-k_{1i}\sigma^2_i - k_{2i}(\sigma^2_i - \mu_{\sigma i})^2) \quad \text{for} \quad k_{2i} > 0,\ V_{\sigma i} < \mu^2_{\sigma i}. \tag{343.4}$$

The product is taken over the l variance components σ^2_i, and k_{1i} and k_{2i} are the parameters of the truncated normal distributions for the individual variance components. To determine these parameters numerically, the results (223.19) to (223.22) may be used.

The covariance components σ_{ij} take on positive and negative values. Because of (223.6) and (343.1) the appropriate choice for the prior is therefore the normal distribution (A11.1). By assuming the k-l covariance components σ_{ij} as being independent we find the prior $p(\sigma_{ij})_{kl}$ for the k-l components σ_{ij} with (A11.3) after neglecting the constants by

$$p(\sigma_{ij})_{kl} \propto \prod_{1}^{k-l} \exp(-(\sigma_{ij} - \mu_{\sigma ij})^2/2V_{\sigma ij}). \tag{343.5}$$

The product is taken over the k-l covariance components σ_{ij}.

By applying the likelihood function (341.7) Bayes' theorem (211.1) now gives the posterior distribution $p(\boldsymbol{\sigma}|\mathbf{y})$ of the vector $\boldsymbol{\sigma}$ of the variance and covariance components σ^2_i and σ_{ij}. If we use the inverted gamma distribution as prior for the variance components σ^2_i, we find with (343.2) and (343.5)

$$p(\boldsymbol{\sigma}|\mathbf{y}) \propto \prod_{i=1}^{l} (1/\sigma^2_i)^{p_i+1} \exp(-b_i/\sigma^2_i) \prod_{1}^{k-l} \exp(-(\sigma_{ij} - \mu_{\sigma ij})^2/2V_{\sigma ij})$$
$$(\det\boldsymbol{\Sigma}\ \det\mathbf{X}'\boldsymbol{\Sigma}^{-1}\mathbf{X})^{-1/2}\exp(-\mathbf{y}'\mathbf{W}\mathbf{y}/2) \tag{343.6}$$

with b_i and p_i from (343.3). If we introduce the truncated normal distribution as prior for the variance components σ^2_i, we obtain with (343.4) and (343.5) the posterior distribution for the vector $\boldsymbol{\sigma}$ of the variance and covariance components σ^2_i and σ_{ij}

$$p(\boldsymbol{\sigma}|\mathbf{y}) \propto \prod_{i=1}^{l} \exp(-k_{1i}\sigma^2_i - k_{2i}(\sigma^2_i - \mu_{\sigma i})^2) \prod_{1}^{k-l} \exp(-(\sigma_{ij} - \mu_{\sigma ij})^2/2V_{\sigma ij})$$
$$(\det\boldsymbol{\Sigma}\ \det\mathbf{X}'\boldsymbol{\Sigma}^{-1}\mathbf{X})^{-1/2}\exp(-\mathbf{y}'\mathbf{W}\mathbf{y}/2) \quad \text{for} \quad k_{2i} > 0,\ V_{\sigma i} < \mu^2_{\sigma i}. \tag{343.7}$$

By integrating over the domain of σ, the normalization factors for the distributions (343.6) and (343.7) are determined. Neither for the distribution (343.6) nor for (343.7) could this integration be solved analytically. The same applies for the integration in (231.5), which gives the Bayes estimate $\hat{\sigma}_B$ for the vector $\boldsymbol{\sigma}$ of variance and covariance components, and for the integration leading to the confidence region (241.2) or to the hypothesis test for the variance and covariance components. To solve these inferential problems, the numerical methods of Section 27 need to be applied.

35 Classification

351 Decision by Bayes Rule

Classification solves the problem of attributing an object to a class based on the characteristics observed for the object. In other words, a sample of the characteristics has been drawn from a population within a set of different populations without knowing from which population. The sample then needs to be attributed to the population from where it stems. Classification has been widely applied in pattern recognition, for instance in automatic reading of numbers or letters.

The predictive analysis of the Bayesian inference is well suited to solve the problem of classification (Geisser 1964; Press 1982, p.402). We will therefore concentrate on this approach. Since more than one characteristic is measured for the object, which needs to be classified, a multivariate analysis is asked for. However, this does not mean that Bayesian inference in a multivariate linear model has to be introduced together with the appropriate distributions, especially the matrix t distribution, see for instance (Koch and Riesmeier 1985). We need only a few results from the multivariate model of the standard statistical techniques, in order to apply the Bayesian predictive analysis for the problem of classification. The Bayesian approach in a multivariate linear model is treated, for instance, in Box and Tiao (1973), Broemeling (1985), Bunke and Bunke (1986), Press (1982), Press (1989) and Zellner (1971).

We assume u different populations ω_i and a p×1 observation vector $\mathbf{y}$ having the normal distribution with the p×1 vector $\boldsymbol{\theta}_i$ and the p×p positive definite matrix $\boldsymbol{\Sigma}_i$ as parameters

$$\mathbf{y}|\boldsymbol{\theta}_i,\boldsymbol{\Sigma}_i \sim N(\boldsymbol{\theta}_i,\boldsymbol{\Sigma}_i) \quad \text{for} \quad i\in\{1,\ldots,u\}, \tag{351.1}$$

if $\mathbf{y}$ originates from the population ω_i. In an application the parameters $\boldsymbol{\theta}_i$ and $\boldsymbol{\Sigma}_i$, which according to (A21.3) are the expected value and the covariance matrix of $\mathbf{y}$, will be usually unknown. We therefore need training samples, which are observations whose origin we know. Thus,

$$\begin{array}{ll} \mathbf{y}_{11}, \mathbf{y}_{12},\ldots, \mathbf{y}_{1,n1} & \text{from } \omega_1 \\ \mathbf{y}_{21}, \mathbf{y}_{22},\ldots, \mathbf{y}_{2,n2} & \text{from } \omega_2 \\ \ldots\ldots\ldots\ldots\ldots\ldots & \\ \mathbf{y}_{u1}, \mathbf{y}_{u2},\ldots, \mathbf{y}_{u,nu} & \text{from } \omega_u , \end{array} \tag{351.2}$$

where the p×1 observation vectors $\mathbf{y}_{ij}$ are independent and normally distributed according to

$$\mathbf{y}_{ij}|\boldsymbol{\theta}_i,\boldsymbol{\Sigma}_i \sim N(\boldsymbol{\theta}_i,\boldsymbol{\Sigma}_i) \quad \text{for} \quad i\in\{1,\dots,u\}, \quad j\in\{1,\dots,n_i\}. \tag{351.3}$$

Let p_i be the prior probability that the observation vector $\mathbf{y}$ stems from the population ω_i and let $p(\mathbf{y}|\mathbf{y}\in\omega_i)$ be the probability density function of the observation vector $\mathbf{y}$ under the condition that $\mathbf{y}$ comes from ω_i. Bayes' theorem (211.1) then gives the posterior density function $p(\mathbf{y}\in\omega_i|\mathbf{y})$ of the observation vector $\mathbf{y}$ being classified into the population ω_i, given the data $\mathbf{y}$,

$$p(\mathbf{y}\in\omega_i|\mathbf{y}) \propto p_i p(\mathbf{y}|\mathbf{y}\in\omega_i). \tag{351.4}$$

The classification is a decision problem. We therefore apply the Bayes rule (231.2) to reach a decision with minimum posterior expected loss. Let $L(\mathbf{y}\in\omega_i,\omega_j)$ denote the loss for the classification of $\mathbf{y}$ into the population ω_i, while actually the population ω_j is present. We assign the values

$$\begin{aligned} L(\mathbf{y}\in\omega_i,\omega_j) &= 0 \quad \text{for} \quad i = j \\ L(\mathbf{y}\in\omega_i,\omega_j) &\neq 0 \quad \text{for} \quad i \neq j, \end{aligned} \tag{351.5}$$

that is no loss for the correct classification and losses different from zero for the misclassification.

The posterior expected loss of the classification $\mathbf{y}$ into ω_i is given with (351.4) and (351.5) by

$$E(L(\mathbf{y}\in\omega_i)) = \sum_{\substack{j=1\\j\neq i}}^{u} L(\mathbf{y}\in\omega_i,\omega_j)p(\mathbf{y}\in\omega_j|\mathbf{y}). \tag{351.6}$$

By minimizing the expected loss, we obtain the Bayes rule for the classification, which says: classify the observation vector $\mathbf{y}$ into the population ω_i, if

$$\sum_{\substack{j=1\\j\neq i}}^{u} L(\mathbf{y}\in\omega_i,\omega_j)p(\mathbf{y}\in\omega_j|\mathbf{y}) < \sum_{\substack{j=1\\j\neq k}}^{u} L(\mathbf{y}\in\omega_k,\omega_j)p(\mathbf{y}\in\omega_j|\mathbf{y}) \quad \text{for all } k\in\{1,\dots,u\}. \tag{351.7}$$

If we assign equal losses to the misclassifications, that is

$$L(\mathbf{y}\in\omega_i,\omega_j) = L(\mathbf{y}\in\omega_k,\omega_l), \tag{351.8}$$

then (351.7) simplifies to the Bayes rule: classify the observation vector $\mathbf{y}$ into the population ω_i, if

$$p(\mathbf{y}\in\omega_i|\mathbf{y}) > p(\mathbf{y}\in\omega_k|\mathbf{y}) \quad \text{for all} \quad k\in\{1,\dots,u\} \quad \text{with} \quad i \neq k. \tag{351.9}$$

In other words, assign the observation vector $\mathbf{y}$ to the population ω_i, for which the posterior density function $p(\mathbf{y}\in\omega_i|\mathbf{y})$ attains a maximum.

So far we have not mentioned how the density function $p(\mathbf{y}\in\omega_i|\mathbf{y})$ in (351.7) or (351.9) is determined. It can be readily written down, if the parameters of the normal distribution (351.1) are known. Although this would be an exception for an application, we start with this case for a comparison with the results, for which the parameters of the distribution (351.1) are unknown.

352 Known Parameters

If the parameters of the normal distributions of the different populations are known, the probability density function $p(\mathbf{y}|\mathbf{y}\in\omega_i)$ of the observation vector $\mathbf{y}$ under the condition that $\mathbf{y}$ comes from ω_i is given by (351.1)

$$p(\mathbf{y}|\mathbf{y}\in\omega_i) = p(\mathbf{y}|\boldsymbol{\theta}_i,\boldsymbol{\Sigma}_i)$$

and we obtain with (A21.1)

$$p(\mathbf{y}|\mathbf{y}\in\omega_i) \propto (\det\boldsymbol{\Sigma}_i)^{-1/2} \exp[-\frac{1}{2}(\mathbf{y}-\boldsymbol{\theta}_i)'\boldsymbol{\Sigma}_i^{-1}(\mathbf{y}-\boldsymbol{\theta}_i)]. \quad (352.1)$$

Substituting this expression in (351.4) gives the posterior density function $p(\mathbf{y}\in\omega_i|\mathbf{y})$ for classifying $\mathbf{y}$ into ω_i, given the data $\mathbf{y}$,

$$p(\mathbf{y}\in\omega_i|\mathbf{y}) \propto p_i(\det\boldsymbol{\Sigma}_i)^{-1/2}\exp[-\frac{1}{2}(\mathbf{y}-\boldsymbol{\theta}_i)'\boldsymbol{\Sigma}_i^{-1}(\mathbf{y}-\boldsymbol{\theta}_i)]. \quad (352.2)$$

The Bayes rule for the classification then follows from (351.7) or from (351.9), if equal losses are assigned to the misclassification.

The latter case will be investigated further by introducing a *discriminant function* $d_i(\mathbf{y})$, which is obtained by taking the natural logarithm of the density (352.2). The Bayes rule then says: classify $\mathbf{y}$ into ω_i, if

$$d_i(\mathbf{y}) > d_j(\mathbf{y}) \quad \text{for all} \quad j\in\{1,\ldots,u\} \quad \text{with} \quad i \neq j \quad (352.3)$$

and

$$d_i(\mathbf{y}) = -\frac{1}{2}(\mathbf{y}-\boldsymbol{\theta}_i)'\boldsymbol{\Sigma}_i^{-1}(\mathbf{y}-\boldsymbol{\theta}_i) - \frac{1}{2}\ln\det\boldsymbol{\Sigma}_i + \ln p_i. \quad (352.4)$$

If $\boldsymbol{\Sigma}_i=\boldsymbol{\Sigma}$ and $p_i=c$ holds for all $i\in\{1,\ldots,u\}$, we omit constants, which is admissible because of (352.3), and obtain instead of (352.4) the negative discriminant function

$$-d_i(\mathbf{y}) = (\mathbf{y}-\boldsymbol{\theta}_i)'\boldsymbol{\Sigma}^{-1}(\mathbf{y}-\boldsymbol{\theta}_i). \quad (352.5)$$

It is known as the *Mahalanobis distance* between the observation vector $\mathbf{y}$ and the population ω_i. This discriminant function classifies the vector $\mathbf{y}$ into the population ω_i from which $\mathbf{y}$ has the shortest Mahalanobis distance. If in addition $\boldsymbol{\Sigma}=\sigma^2\mathbf{I}$ we find again after omitting the constant the negative discriminant function

$$-d_i(\mathbf{y}) = (\mathbf{y}-\boldsymbol{\theta}_i)'(\mathbf{y}-\boldsymbol{\theta}_i), \qquad (352.6)$$

which is known as *minimum distance classifier*.

While the discriminant function (352.4) depends on the quadratic form $\mathbf{y}'\boldsymbol{\Sigma}_i^{-1}\mathbf{y}$ of the observation vector $\mathbf{y}$, the discriminant functions (352.5) and (352.6) are linear in $\mathbf{y}$, since $\mathbf{y}'\boldsymbol{\Sigma}^{-1}\mathbf{y}$ and $\mathbf{y}'\mathbf{y}$ are constants. The boundaries determined by (352.5) or (352.6) between different populations are therefore hyperplanes, while the boundaries resulting from (352.4) are surfaces of second order.

353 Unknown Parameters

We will now derive the density function $p(\mathbf{y}|\mathbf{y}\epsilon\omega_i)$ in (351.4) under the assumption that the parameters $\boldsymbol{\theta}_i$ and $\boldsymbol{\Sigma}_i$ of the normal distribution for the population ω_i are unknown. This is typical for a practical application of the classification.

We will later have to integrate with respect to $\boldsymbol{\theta}_i$ and $\boldsymbol{\Sigma}_i$. It is therefore convenient to introduce instead of the covariance matrix $\boldsymbol{\Sigma}_i$ as parameter the weight matrix $\mathbf{P}_i$ with

$$\mathbf{P}_i = \boldsymbol{\Sigma}_i^{-1}. \qquad (353.1)$$

Given are the training samples (351.2), from which we derive the unbiased estimate $\hat{\boldsymbol{\theta}}_i$ of $\boldsymbol{\theta}_i$ and $\hat{\boldsymbol{\Sigma}}_i$ of $\boldsymbol{\Sigma}_i$ with (Koch 1988a, p.291)

$$\hat{\boldsymbol{\theta}}_i = \frac{1}{n_i}\sum_{j=1}^{n_i} \mathbf{y}_{ij} \qquad (353.2)$$

and

$$\hat{\boldsymbol{\Sigma}}_i = \frac{1}{n_i-1}\boldsymbol{\Omega}_i \qquad (353.3)$$

with

$$\boldsymbol{\Omega}_i = \sum_{j=1}^{n_i} (\mathbf{y}_{ij}-\hat{\boldsymbol{\theta}}_i)(\mathbf{y}_{ij}-\hat{\boldsymbol{\theta}}_i)'.$$

Since the observation vectors $\mathbf{y}_{ij}$ are independent and normally distributed according to (351.3), the estimate $\hat{\boldsymbol{\theta}}_i$ under the condition that $\boldsymbol{\theta}_i$ and $\boldsymbol{\Sigma}_i$ are given is normally distributed (Koch 1988a, p.143)

$$\hat{\boldsymbol{\theta}}_i | \boldsymbol{\theta}_i, \boldsymbol{\Sigma}_i \sim N(\boldsymbol{\theta}_i, n_i^{-1}\boldsymbol{\Sigma}_i). \tag{353.4}$$

With (353.1) and (A21.1) we therefore obtain the density function $p(\hat{\boldsymbol{\theta}}_i | \boldsymbol{\theta}_i, \mathbf{P}_i^{-1})$ of $\hat{\boldsymbol{\theta}}_i$ with

$$p(\hat{\boldsymbol{\theta}}_i | \boldsymbol{\theta}_i, \mathbf{P}_i^{-1}) \propto (\det \mathbf{P}_i)^{1/2} \exp[-\frac{1}{2} n_i (\hat{\boldsymbol{\theta}}_i - \boldsymbol{\theta}_i)' \mathbf{P}_i (\hat{\boldsymbol{\theta}}_i - \boldsymbol{\theta}_i)]. \tag{353.5}$$

The matrix $\boldsymbol{\Omega}_i$ in (353.3) has the Wishart distribution (Koch 1988a, p.302)

$$\boldsymbol{\Omega}_i | \boldsymbol{\theta}_i, \boldsymbol{\Sigma}_i \sim W(n_i - 1, \boldsymbol{\Sigma}_i). \tag{353.6}$$

The density function $p(\boldsymbol{\Omega}_i | \boldsymbol{\theta}_i, \boldsymbol{\Sigma}_i)$ of the matrix $\boldsymbol{\Omega}_i$ therefore follows with the substitution (353.1) by (Koch 1988a, p.160)

$$p(\boldsymbol{\Omega}_i | \boldsymbol{\theta}_i, \mathbf{P}_i^{-1}) \propto (\det \mathbf{P}_i)^{(n_i-1)/2} (\det \boldsymbol{\Omega}_i)^{(n_i-p-2)/2} \exp[-\frac{1}{2} \operatorname{tr}(\mathbf{P}_i \boldsymbol{\Omega}_i)]. \tag{353.7}$$

Now we are in the position to derive the posterior distribution for the parameters $\boldsymbol{\theta}_i$ and $\mathbf{P}_i$. Let $p(\boldsymbol{\theta}_i, \mathbf{P}_i)$ be the prior density function of $\boldsymbol{\theta}_i$ and $\mathbf{P}_i$ and $p(\hat{\boldsymbol{\theta}}_i, \boldsymbol{\Omega}_i | \boldsymbol{\theta}_i, \mathbf{P}_i)$ the likelihood function of the estimates $\hat{\boldsymbol{\theta}}_i$ and $\boldsymbol{\Omega}_i$ determined by the training sample. The posterior density function $p(\boldsymbol{\theta}_i, \mathbf{P}_i | \hat{\boldsymbol{\theta}}_i, \boldsymbol{\Omega}_i)$ of $\boldsymbol{\theta}_i$ and $\mathbf{P}_i$ given the estimates $\hat{\boldsymbol{\theta}}_i$ and $\boldsymbol{\Omega}_i$ then follows with Bayes' theorem (211.1) by

$$p(\boldsymbol{\theta}_i, \mathbf{P}_i | \hat{\boldsymbol{\theta}}_i, \boldsymbol{\Omega}_i) \propto p(\boldsymbol{\theta}_i, \mathbf{P}_i) p(\hat{\boldsymbol{\theta}}_i, \boldsymbol{\Omega}_i | \boldsymbol{\theta}_i, \mathbf{P}_i). \tag{353.8}$$

For introducing the prior distribution $p(\boldsymbol{\theta}_i, \mathbf{P}_i)$ we have the choice between a noninformative prior and an informative prior. With respect to the practical application, where in general nothing is known in advance about the parameters $\boldsymbol{\theta}_i$ and $\mathbf{P}_i$ of the different populations, we will restrict ourselves to noninformative priors. Informative priors may be introduced as conjugate priors. The normal-Wishart distribution used in the multivariate linear model (Broemeling 1985, p.377; Bunke and Bunke 1986, p.439) would be such a prior.

Based on Jeffrey's principle of invariance the noninformative prior $p(\boldsymbol{\theta}_i, \mathbf{P}_i)$ for $\boldsymbol{\theta}_i$ and $\mathbf{P}_i$ follows from (222.10) with (Geisser 1965; Press 1982, p.80)

$$p(\boldsymbol{\theta}_i, \mathbf{P}_i) \propto (\det \mathbf{P}_i)^{-(p+1)/2}. \tag{353.9}$$

Since $\hat{\boldsymbol{\theta}}_i$ and $\boldsymbol{\Omega}_i$ are independent, we obtain with (353.5), (353.7) and (353.9) instead of (353.8)

$$p(\boldsymbol{\theta}_i, \mathbf{P}_i | \hat{\boldsymbol{\theta}}_i, \boldsymbol{\Omega}_i) \propto (\det \mathbf{P}_i)^{(n_i-p-1)/2}$$
$$\exp\{-\frac{1}{2} \operatorname{tr}[\mathbf{P}_i(\boldsymbol{\Omega}_i + n_i(\hat{\boldsymbol{\theta}}_i - \boldsymbol{\theta}_i)(\hat{\boldsymbol{\theta}}_i - \boldsymbol{\theta}_i)')]\}. \tag{353.10}$$

Now we compute the predictive density function $p(\mathbf{y}|\hat{\boldsymbol{\theta}}_i, \boldsymbol{\Omega}_i)$ of the predicted observation vector $\mathbf{y}$ under the condition that the estimates $\hat{\boldsymbol{\theta}}_i$ and $\boldsymbol{\Omega}_i$ are given, which means that the observed data belong to the population ω_i. This density is obtained with (262.2) by

$$p(\mathbf{y}|\hat{\boldsymbol{\theta}}_i,\boldsymbol{\Omega}_i) \propto \int_{\Theta_i}\int_{P_i} p(\mathbf{y}|\boldsymbol{\theta}_i,\mathbf{P}_i,\hat{\boldsymbol{\theta}}_i,\boldsymbol{\Omega}_i)p(\boldsymbol{\theta}_i,\mathbf{P}_i|\hat{\boldsymbol{\theta}}_i,\boldsymbol{\Omega}_i)d\mathbf{P}_i d\boldsymbol{\theta}_i\,, \tag{353.11}$$

where Θ_i denotes the parameter space of $\boldsymbol{\theta}_i$ and P_i the parameter space of $\mathbf{P}_i$. For the density $p(\mathbf{y}|\boldsymbol{\theta}_i,\mathbf{P}_i,\hat{\boldsymbol{\theta}}_i,\boldsymbol{\Omega}_i)$ of the predicted observation vector $\mathbf{y}$ we assume the normal distribution (351.1) together with (353.1). Thus, with (A21.1)

$$p(\mathbf{y}|\boldsymbol{\theta}_i,\mathbf{P}_i,\hat{\boldsymbol{\theta}}_i,\boldsymbol{\Omega}_i) \propto (\det\mathbf{P}_i)^{1/2}\exp[-\frac{1}{2}(\mathbf{y}-\boldsymbol{\theta}_i)'\mathbf{P}_i(\mathbf{y}-\boldsymbol{\theta}_i)]\,. \tag{353.12}$$

Substituting (353.10) and (353.12) in (353.11) leads to

$$p(\mathbf{y}|\hat{\boldsymbol{\theta}}_i,\boldsymbol{\Omega}_i) \propto \int_{\Theta_i}\int_{P_i} (\det\mathbf{P}_i)^{(n_i-p)/2}\exp[-\frac{1}{2}\,\mathrm{tr}(\mathbf{P}_i\mathbf{A})]d\mathbf{P}_i d\boldsymbol{\theta}_i \tag{353.13}$$

with

$$\mathbf{A} = \boldsymbol{\Omega}_i + n_i(\hat{\boldsymbol{\theta}}_i-\boldsymbol{\theta}_i)(\hat{\boldsymbol{\theta}}_i-\boldsymbol{\theta}_i)' + (\mathbf{y}-\boldsymbol{\theta}_i)(\mathbf{y}-\boldsymbol{\theta}_i)'\,.$$

The integrand is proportional to the density of the Wishart distribution for the matrix $\mathbf{P}_i$ with n_i+1 and $\mathbf{A}$ as parameters. Because of (211.6) the integration with respect to $\mathbf{P}_i$ therefore gives

$$p(\mathbf{y}|\hat{\boldsymbol{\theta}}_i,\boldsymbol{\Omega}_i) \propto \int_{\Theta_i} (\det\mathbf{A})^{-(n_i+1)/2}d\boldsymbol{\theta}_i\,. \tag{353.14}$$

For the integration with respect to $\boldsymbol{\theta}_i$ we complete the squares on $\boldsymbol{\theta}_i$ and obtain

$$\mathbf{A} = (n_i+1)\boldsymbol{\theta}_i\boldsymbol{\theta}_i' - \boldsymbol{\theta}_i(n_i\hat{\boldsymbol{\theta}}_i'+\mathbf{y}') - (n_i\hat{\boldsymbol{\theta}}_i+\mathbf{y})\boldsymbol{\theta}_i' + \boldsymbol{\Omega}_i + n_i\hat{\boldsymbol{\theta}}_i\hat{\boldsymbol{\theta}}_i' + \mathbf{y}\mathbf{y}'$$

or

$$(n_i+1)^{-1}\mathbf{A} = (\boldsymbol{\theta}_i-\bar{\boldsymbol{\theta}}_i)(\boldsymbol{\theta}_i-\bar{\boldsymbol{\theta}}_i)' + \mathbf{B} \tag{353.15}$$

with

$$\bar{\boldsymbol{\theta}}_i = (n_i+1)^{-1}(n_i\hat{\boldsymbol{\theta}}_i+\mathbf{y})$$

and

$$\mathbf{B} = (n_i+1)^{-1}(\boldsymbol{\Omega}_i+n_i\hat{\boldsymbol{\theta}}_i\hat{\boldsymbol{\theta}}_i'+\mathbf{y}\mathbf{y}') - (n_i+1)^{-2}(n_i\hat{\boldsymbol{\theta}}_i+\mathbf{y})(n_i\hat{\boldsymbol{\theta}}_i+\mathbf{y})'\,.$$

Substituting (353.15) in (353.14) and omitting the constants gives

$$p(\mathbf{y}|\hat{\boldsymbol{\theta}}_i,\boldsymbol{\Omega}_i) \propto \int_{\Theta_i} [\det((\boldsymbol{\theta}_i-\bar{\boldsymbol{\theta}}_i)(\boldsymbol{\theta}_i-\bar{\boldsymbol{\theta}}_i)'+\mathbf{B})]^{-(n_i+1)/2} d\boldsymbol{\theta}_i$$

$$\propto (\det\mathbf{B})^{-(n_i+1)/2} \int_{\Theta_i} [\det(\mathbf{I}+\mathbf{B}^{-1}(\boldsymbol{\theta}_i-\bar{\boldsymbol{\theta}}_i)(\boldsymbol{\theta}_i-\bar{\boldsymbol{\theta}}_i)')]^{-(n_i+1)/2} d\boldsymbol{\theta}_i . \quad (353.16)$$

Using the identity (Press 1982, p.20)

$$\det(\mathbf{I}_l+\mathbf{CD}) = \det(\mathbf{I}_m+\mathbf{DC}), \quad (353.17)$$

where $\mathbf{C}$ is an $l\times m$ and $\mathbf{D}$ an $m\times l$ matrix, we obtain

$$p(\mathbf{y}|\hat{\boldsymbol{\theta}}_i,\boldsymbol{\Omega}_i) \propto (\det\mathbf{B})^{-(n_i+1)/2} \int_{\Theta_i} (1+(\boldsymbol{\theta}_i-\bar{\boldsymbol{\theta}}_i)'\mathbf{B}^{-1}(\boldsymbol{\theta}_i-\bar{\boldsymbol{\theta}}_i))^{-(n_i+1)/2} d\boldsymbol{\theta}_i . \quad (353.18)$$

By comparing the integrand with (A22.3) we recognize $\boldsymbol{\theta}_i$ as having the multivariate t-distribution. The integration with respect to $\boldsymbol{\theta}_i$ therefore gives

$$p(\mathbf{y}|\hat{\boldsymbol{\theta}}_i,\boldsymbol{\Omega}_i) \propto (\det\mathbf{B})^{-(n_i+1)/2}(\det\mathbf{B})^{1/2} = (\det\mathbf{B})^{-n_i/2}. \quad (353.19)$$

With $\mathbf{B}$ from (353.15) we find

$$(n_i+1)\mathbf{B} = \boldsymbol{\Omega}_i + \frac{n_i}{n_i+1}\,[(n_i+1)\hat{\boldsymbol{\theta}}_i\hat{\boldsymbol{\theta}}_i' + \frac{n_i+1}{n_i}\,\mathbf{y}\mathbf{y}' - \frac{1}{n_i}\,(n_i\hat{\boldsymbol{\theta}}_i+\mathbf{y})(n_i\hat{\boldsymbol{\theta}}_i+\mathbf{y})']$$

$$= \boldsymbol{\Omega}_i + \frac{n_i}{n_i+1}\,(\mathbf{y}-\hat{\boldsymbol{\theta}}_i)(\mathbf{y}-\hat{\boldsymbol{\theta}}_i)' \quad (353.20)$$

and

$$p(\mathbf{y}|\hat{\boldsymbol{\theta}}_i,\boldsymbol{\Omega}_i) \propto [\det(\boldsymbol{\Omega}_i + \frac{n_i}{n_i+1}\,(\mathbf{y}-\hat{\boldsymbol{\theta}}_i)(\mathbf{y}-\hat{\boldsymbol{\theta}}_i)')]^{-n_i/2}$$

$$\propto [\det(\mathbf{I} + \frac{n_i}{n_i+1}\,\boldsymbol{\Omega}_i^{-1}(\mathbf{y}-\hat{\boldsymbol{\theta}}_i)(\mathbf{y}-\hat{\boldsymbol{\theta}}_i)')]^{-n_i/2}. \quad (353.21)$$

By applying the identity (353.17) again we obtain with (353.3)

$$p(\mathbf{y}|\hat{\boldsymbol{\theta}}_i,\hat{\boldsymbol{\Sigma}}_i) \propto (1 + \frac{n_i}{n_i^2-1}\,(\mathbf{y}-\hat{\boldsymbol{\theta}}_i)'\hat{\boldsymbol{\Sigma}}_i^{-1}(\mathbf{y}-\hat{\boldsymbol{\theta}}_i))^{-n_i/2}. \quad (353.22)$$

A comparison with (A22.3) reveals that the predictive density stems from the multivariate t-distribution. Thus,

$$\mathbf{y}|\hat{\boldsymbol{\theta}}_i,\hat{\boldsymbol{\Sigma}}_i \sim t(\hat{\boldsymbol{\theta}}_i,(n_i^2-1)\hat{\boldsymbol{\Sigma}}_i/(n_i(n_i-p)),n_i-p) \quad (353.23)$$

and

$$p(\mathbf{y}|\hat{\boldsymbol{\theta}}_i,\hat{\boldsymbol{\Sigma}}_i) \propto \frac{(n_i)^{p/2}\Gamma(n_i/2)(\det\hat{\boldsymbol{\Sigma}}_i)^{-1/2}}{(n_i^2-1)^{p/2}\Gamma((n_i-p)/2)}$$

$$(1 + \frac{n_i}{n_i^2-1}(\mathbf{y}-\hat{\boldsymbol{\theta}}_i)'\hat{\boldsymbol{\Sigma}}_i^{-1}(\mathbf{y}-\hat{\boldsymbol{\theta}}_i))^{-n_i/2}. \quad (353.24)$$

The predictive density function $p(\mathbf{y}|\hat{\boldsymbol{\theta}}_i,\hat{\boldsymbol{\Sigma}}_i)$ of the predicted observation vector $\mathbf{y}$ under the condition that the observed data belong to the population ω_i may be denoted by $p(\mathbf{y}|\mathbf{y}\in\omega_i)$. In analogy to (351.4) we therefore obtain the predictive density function $p(\mathbf{y}\in\omega_i|\mathbf{y})$ of the observation vector $\mathbf{y}$ being classified into the population ω_i, given the data $\mathbf{y}$,

$$p(\mathbf{y}\in\omega_i|\mathbf{y}) \propto \frac{(n_i)^{p/2}\Gamma(n_i/2)(\det\hat{\boldsymbol{\Sigma}}_i)^{-1/2}p_i}{(n_i^2-1)^{p/2}\Gamma((n_i-p)/2)}$$

$$(1 + \frac{n_i}{n_i^2-1}(\mathbf{y}-\hat{\boldsymbol{\theta}}_i)'\hat{\boldsymbol{\Sigma}}_i^{-1}(\mathbf{y}-\hat{\boldsymbol{\theta}}_i))^{-n_i/2}. \quad (353.25)$$

The Bayes rule for classifying $\mathbf{y}$ into ω_i now follows from (351.7) or from (351.9) for equal losses of the misclassifications.

We compare the density (353.25) for classifying $\mathbf{y}$ into ω_i with the result (352.2), which is valid for known parameters $\boldsymbol{\theta}_i$ and $\boldsymbol{\Sigma}_i$. We recognize that these parameters, in case they are unknown, should not be just replaced by their estimates $\hat{\boldsymbol{\theta}}_i$ and $\hat{\boldsymbol{\Sigma}}_i$. As (353.25) shows, the density for the classification should also account for different sizes n_i of the training samples.

With

$$n_i = n \quad \text{for} \quad i\in\{1,\ldots,u\} \quad (353.26)$$

we find instead of (353.25)

$$p(\mathbf{y}\in\omega_i|\mathbf{y}) \propto p_i(\det\hat{\boldsymbol{\Sigma}}_i)^{-1/2}(1 + \frac{n}{n^2-1}(\mathbf{y}-\hat{\boldsymbol{\theta}}_i)'\hat{\boldsymbol{\Sigma}}_i^{-1}(\mathbf{y}-\hat{\boldsymbol{\theta}}_i))^{-n/2}. \quad (353.27)$$

This result is very similar to (352.2) after substituting $\boldsymbol{\theta}_i$ and $\boldsymbol{\Sigma}_i$ by the estimates $\hat{\boldsymbol{\theta}}_i$ and $\hat{\boldsymbol{\Sigma}}_i$.

Example 1: Digital images of the surface of the earth are recorded by special satellites. Multispectral scanners on board the satellites decompose the images of the surface into picture elements, so-called pixels, and measure the gray levels of the pixels in different bands of the frequency of electromagnetic waves. If there are p sensors for the different frequency bands, the data vector $\mathbf{y}$ of each pixel contains p different gray levels.

Let the data vector $\mathbf{y}$ of a pixel be classified into u different populations ω_i. We select training sites for each population. Different numbers of pixels are available within each training site, so that the observations (351.2) are collected. We assume equal losses for the misclassification. The Bayes rule for classifying the data vector $\mathbf{y}$ of a pixel into the population ω_i then follows from (351.9) with the posterior density function $p(\mathbf{y} \in \omega_i | \mathbf{y})$ for the classification from (353.25). Δ

36 Posterior Analysis Based on Distributions for Robust Maximum Likelihood Type Estimates

361 Introduction

In the recent past considerable attention has been paid to robust statistics. Robustness means insensitivity to small deviations from the underlying assumptions. In Bayesian analysis one studies, for instance, the robustness of the posterior distribution with respect to possible misspecifications of the prior density function (Berger 1985, p.195).

For the standard statistical techniques much effort has been spent to derive robust estimates (Hampel et al. 1986; Hoaglin et al. 1983; Huber 1981). The reason is that frequently deviations occur from the distributions assumed for the data. Because of a few outliers the observations are not normally distributed, as usually assumed, but obey a distribution which slightly differs from the normal distribution. In geodesy robust estimation has therefore been mostly applied to detect outliers or to diminish the effect of outliers on the estimation of parameters, but also to detect moving points in a deformation analysis (Caspary 1988; Jorgenson et al. 1985; Somogyi 1988).

Huber's robust estimate is recommended, when the distribution of the data is close to the normal distribution, while for samples very badly contaminated by outliers the robust redescending estimate, for instance, of Hampel's type should be applied (Hoaglin et al. 1983, p.397). Both these estimates are maximum likelihood type estimates, so-called M-estimates. Distributions therefore exist, which, assumed for the data, yield in a maximum likelihood estimation the robust estimates.

Maximum likelihood estimates are also obtained by the Bayesian analysis as shown with (232.4), so that the robust M-estimates can be also derived by Bayesian inference. We only have to introduce the distributions leading to the M-estimates as likelihood functions and to assume noninformative priors for the unknown parameters, in order to obtain the posterior distribution for the parameters which gives as a maximum likelihood estimate the robust M-estimate. But by means of the posterior distribution we not only find the M-estimate, but we may also solve any inferential task for the parameters. For instance, we can apply the predictive analysis in order to decide which observations are outliers.

362 Likelihood Function

Let the n observations y_i, collected in the observation vector $\mathbf{y}=(y_i)$, be independent and have equal variances σ^2. By starting from the nonlinear model (321.1) we have with $\mathbf{f}(\boldsymbol{\beta})=(f_i(\boldsymbol{\beta}))$, where $\mathbf{f}$ denotes an n×1 given vector of nonlinear functions of the u×1 vector $\boldsymbol{\beta}$ of unknown parameters,

$$E(y_i|\boldsymbol{\beta}) = f_i(\boldsymbol{\beta}) \quad \text{with} \quad V(y_i|\sigma^2) = \sigma^2 \quad \text{for} \quad i\in\{1,\ldots,n\}. \tag{362.1}$$

In the case of a linear model, different weights may be given for the observations y_i. The transformation (311.4) reduces such a model to (362.1). However, a weight matrix $\mathbf{P}$ for the observations, which is not a diagonal matrix, cannot be handled, since outliers in the original and not in linearly transformed observations shall be accounted for by the robust estimation.

We assume distributions of the following type for the observations y_i

$$p(y_i|\boldsymbol{\beta},\sigma) \propto \frac{1}{\sigma}\, g(x), \tag{362.2}$$

where σ denotes the standard deviation of the observations with $\sigma=\sqrt{\sigma^2}$ and

$$x = (y_i-f_i(\boldsymbol{\beta}))/\sigma.$$

As can be seen in this expression, $\boldsymbol{\beta}$ determines the location of the graph of the density function, while σ determines its scale. Especially in the context of robust estimation $\boldsymbol{\beta}$ is therefore referred to as *location* parameter and σ as *scale* parameter.

By substituting

$$g(x) = \exp(-\rho(x)) \tag{362.3}$$

the distribution (362.2) includes

a) the normal distribution (A11.1) with

$$\rho(x) = x^2/2, \tag{362.4}$$

b) the least informative distribution for location (Huber 1981, p.71)

$$\begin{aligned} \rho(x) &= x^2/2 && \text{for} \quad |x| \le c \\ \rho(x) &= c|x| - c^2/2 && \text{for} \quad |x| > c. \end{aligned} \tag{362.5}$$

This distribution leads to Huber's robust M-estimate. c is a constant, it depends on the contamination of the observations by outliers. From studying empirical distributions of large samples (Huber 1981, p.87, 91) a choice of c=1.5 seems to be quite reasonable.

c) the distribution (Hoaglin et al. 1983, p.367)

$$\rho(x) = x^2/2 \qquad \text{for} \quad |x| \le a$$

$$\rho(x) = a|x| - a^2/2 \qquad \text{for} \quad a < |x| \le b$$

$$\rho(x) = ab - \frac{a^2}{2} + (c-b)\,\frac{a}{2}[1-(\frac{c-|x|}{c-b})^2] \qquad \text{for} \quad b < |x| \le c$$

$$\rho(x) = ab - a^2/2 + (c-b)a/2 \qquad \text{for} \quad |x| > c. \tag{362.6}$$

This distribution gives the redescending M-estimate of Hampel. a, b, c are constants, for instance, a=2, b=4, c=8.

The observations y_i were assumed as being independent, the distribution (362.2) for y_i therefore gives the likelihood function $p(\mathbf{y}|\boldsymbol{\beta},\sigma)$ by

$$p(\mathbf{y}|\boldsymbol{\beta},\sigma) \propto \frac{1}{\sigma^n} \prod_{i=1}^{n} g(\frac{y_i - f_i(\boldsymbol{\beta})}{\sigma}). \tag{362.7}$$

363 Posterior Distribution in the Case of Known Scale

We will assume that the standard deviation σ of the observations is known in the likelihood function (362.7). We will also suppose a noninformative prior for the parameter vector $\boldsymbol{\beta}$. The likelihood function (362.7) is data translated, which means the data y_i only change the location of the graph of the density function. As already discussed in the Example 1 Section 222, the appropriate noninformative prior for the parameter vector $\boldsymbol{\beta}$ is therefore

$$p(\boldsymbol{\beta}) \propto \text{const}. \tag{363.1}$$

Bayes' theorem (211.1) leads with the likelihood function (362.7) to the posterior density function $p(\boldsymbol{\beta}|\mathbf{y})$ for $\boldsymbol{\beta}$

$$p(\boldsymbol{\beta}|\mathbf{y}) \propto \prod_{i=1}^{n} g(\frac{y_i - f_i(\boldsymbol{\beta})}{\sigma}). \tag{363.2}$$

All inferential tasks for the parameter vector $\boldsymbol{\beta}$ can be solved by this distribution.

First we want to estimate $\boldsymbol{\beta}$ by means of the maximum likelihood estimate (232.4). We therefore take the natural logarithm of (363.2)

$$\ln p(\boldsymbol{\beta}|\mathbf{y}) \propto \sum_{i=1}^{n} \ln g(\frac{y_i - f_i(\boldsymbol{\beta})}{\sigma})$$

and set the derivative with respect to β_j in $\boldsymbol{\beta}=(\beta_j)$ equal to zero. This is admissible because of

$$\partial \ln p(\boldsymbol{\beta}|\mathbf{y})/\partial\beta_j = (1/p(\boldsymbol{\beta}|\mathbf{y}))\partial p(\boldsymbol{\beta}|\mathbf{y})/\partial\beta_j = 0$$

so that $\partial p(\boldsymbol{\beta}|\mathbf{y})/\partial\beta_j=0$ is obtained. Thus,

$$\frac{\partial \ln p(\boldsymbol{\beta}|\mathbf{y})}{\partial\beta_j} \propto -\frac{1}{\sigma}\sum_{i=1}^{n}\frac{g'[(y_i-f_i(\boldsymbol{\beta}))/\sigma]}{g[(y_i-f_i(\boldsymbol{\beta}))/\sigma]}\frac{\partial f_i(\boldsymbol{\beta})}{\partial\beta_j} = 0,$$

which leads to

$$\sum_{i=1}^{n}\psi\Big(\frac{y_i-f_i(\boldsymbol{\beta})}{\sigma}\Big)\frac{\partial f_i(\boldsymbol{\beta})}{\partial\beta_j} = 0 \quad \text{for} \quad j \in \{1,\dots,u\} \qquad (363.3)$$

with

$$\psi(x) = -\frac{g'(x)}{g(x)}\,.$$

These equations have to be solved for the maximum likelihood estimate of the parameter vector $\boldsymbol{\beta}$.

By substituting (362.3) in (363.3) we obtain

$$\psi(x) = \rho'(x).$$

Depending on the choice of the distributions introduced in Section 362 we have

a) for the normal distribution (362.4)

$$\psi(x) = x. \qquad (363.4)$$

The equations to be solved according to (363.3) are

$$\sum_{i=1}^{n}(y_i-f_i(\boldsymbol{\beta}))\frac{\partial f_i(\boldsymbol{\beta})}{\partial\beta_j} = 0,$$

which represent the least squares solution of the nonlinear model (362.1).

For a linear model we have

$$\mathbf{f}(\boldsymbol{\beta}) = \mathbf{X}\boldsymbol{\beta} \quad \text{with} \quad \mathbf{X} = (x_{ij})$$

and

$$\frac{\partial f_i(\boldsymbol{\beta})}{\partial\beta_j} = x_{ij}.$$

This gives with $\hat{\boldsymbol{\beta}}$ being the estimate of $\boldsymbol{\beta}$

$$\sum_{i=1}^{n}(y_i-(\mathbf{X}\hat{\boldsymbol{\beta}})_i)x_{ij} = 0$$

or

$$\mathbf{X}'\mathbf{X}\hat{\boldsymbol{\beta}} = \mathbf{X}'\mathbf{y},$$

which are the well-known normal equations, already derived with (312.11).

b) for the least informative distribution (362.5)

$$\begin{aligned} \psi(x) &= x && \text{for} \quad |x| \le c \\ \psi(x) &= c \,\operatorname{sign}(x) && \text{for} \quad |x| > c. \end{aligned} \tag{363.5}$$

This choice of the function $\psi(x)$ leads to Huber's robust M-estimate (Huber 1977).

c) for the distribution (362.6)

$$\begin{aligned} \psi(x) &= x && \text{for} \quad |x| \le a \\ \psi(x) &= a \,\operatorname{sign}(x) && \text{for} \quad a < |x| \le b \\ \psi(x) &= a \frac{x - c\,\operatorname{sign}(x)}{b-c} && \text{for} \quad b < |x| \le c \\ \psi(x) &= 0 && \text{for} \quad |x| > c. \end{aligned} \tag{363.6}$$

This gives the robust redescending M-estimate of Hampel (1974).

For easy reference we give a simple numerical procedure based on modified residuals for computing robust M-estimates (Huber 1981, p.181). We start with a first approximation $\boldsymbol{\beta}^{(1)}$ obtained, for instance, by the method of least squares. In the mth iteration we compute

$$-e_i = y_i - f_i(\boldsymbol{\beta}^{(m)}) \tag{363.7}$$

$$-e_i^* = \psi\left(\frac{-e_i}{\sigma}\right)\sigma \tag{363.8}$$

$$x_{ij} = \frac{\partial}{\partial \beta_j} f_i(\boldsymbol{\beta}^{(m)}), \tag{363.9}$$

where e_i^* denotes the so-called Winsorized residual. Then the normal equations

$$\mathbf{X}'\mathbf{X}\boldsymbol{\tau} = \mathbf{X}'\mathbf{r}^* \tag{363.10}$$

with $\mathbf{X}=(x_{ij})$ and $\mathbf{r}^*=(-e_i^*)$ are solved to obtain $\boldsymbol{\beta}^{(m+1)}$ for the iteration m+1 by

$$\boldsymbol{\beta}^{(m+1)} = \boldsymbol{\beta}^{(m)} + \boldsymbol{\tau}. \tag{363.11}$$

364 Posterior Distribution in the Case of Unknown Scale

We will now assume that the standard deviation σ in the likelihood function (362.7) is unknown. Again we suppose a noninformative prior distribution for the parameter vector $\boldsymbol{\beta}$. Thus, from (363.1) we take

$$p(\boldsymbol{\beta}) \propto \text{const.} \tag{364.1}$$

In order to obtain a posterior density function which leads in a maximum likelihood estimate to the robust estimates of $\boldsymbol{\beta}$ and σ, we introduce in contrast to the noninformative prior to be derived from (222.3) for σ the prior distribution

$$p(\sigma) \propto \text{const.} \tag{364.2}$$

Bayes' theorem (211.1) together with the likelihood function (362.7) gives the posterior density function $p(\boldsymbol{\beta},\sigma|\mathbf{y})$ for $\boldsymbol{\beta}$ and σ

$$p(\boldsymbol{\beta},\sigma|\mathbf{y}) \propto \frac{1}{\sigma^n}\prod_{i=1}^{n} g\left(\frac{y_i-f_i(\boldsymbol{\beta})}{\sigma}\right). \tag{364.3}$$

All inferential problems for the parameter vector $\boldsymbol{\beta}$ and the standard deviation σ of the observations can be solved by this distribution.

We want to estimate $\boldsymbol{\beta}$ and σ by the maximum likelihood estimate (232.4). For $\boldsymbol{\beta}$ we obtain (363.3), as shown in the preceding section. To estimate σ in addition, we take the natural logarithm of (364.3)

$$\ln p(\boldsymbol{\beta},\sigma|\mathbf{y}) \propto \ln\frac{1}{\sigma^n} + \sum_{i=1}^{n} \ln g\left(\frac{y_i-f_i(\boldsymbol{\beta})}{\sigma}\right)$$

and set the derivative with respect to σ equal to zero

$$\frac{\partial \ln p(\boldsymbol{\beta},\sigma|\mathbf{y})}{\partial\sigma} \propto -n\frac{\sigma^n}{\sigma^{n+1}} - \sum_{i=1}^{n}\frac{g'[(y_i-f_i(\boldsymbol{\beta}))/\sigma]}{g[(y_i-f_i(\boldsymbol{\beta}))/\sigma]}\,\frac{y_i-f_i(\boldsymbol{\beta})}{\sigma^2} = 0.$$

We obtain

$$\sum_{i=1}^{n} \chi\left(\frac{y_i-f_i(\boldsymbol{\beta})}{\sigma}\right) = 0 \tag{364.4}$$

with

$$\chi(x) = x\psi(x) - 1.$$

This equation together with (363.3) has to be solved for a robust estimation of the parameter vector $\boldsymbol{\beta}$ and the standard deviation σ in the nonlinear model (362.1) (Huber 1981, p.176).

For a numerical computation of the robust estimate of σ we apply iteratively (Huber 1977)

$$(\sigma^{(m+1)})^2 = \frac{1}{a} \sum_{i=1}^{n} \chi \left(\frac{y_i - f_i(\boldsymbol{\beta}^{(m)})}{\sigma^{(m)}} \right) (\sigma^{(m)})^2 \tag{364.5}$$

with $\sigma^{(m)}$ and $\sigma^{(m+1)}$ being the estimates of σ of the iteration m and m+1, respectively, and with

$$a = (n-u)\, E[(\psi(x))^2], \tag{364.6}$$

where x is a random variable with the standard normal distribution $x \sim N(0,1)$. With $\psi(x)$ from (363.5) for instance, we find

$$E[(\psi(x))^2] = \frac{1}{\sqrt{2\pi}} \left[\int_{-\infty}^{-c} c^2 \exp(-\frac{x^2}{2}) dx + \int_{-c}^{c} x^2 \exp(-\frac{x^2}{2}) dx + \int_{c}^{\infty} c^2 \exp(-\frac{x^2}{2}) dx \right]$$

$$= \frac{1}{\sqrt{2\pi}} \left[2c^2 \int_{c}^{\infty} \exp(-\frac{x^2}{2}) dx + \int_{-c}^{c} x^2 \exp(-\frac{x^2}{2}) dx \right]. \tag{364.7}$$

Since σ is estimated in addition, the iterative procedure for estimating $\boldsymbol{\beta}$ has to be modified by replacing (363.8) with

$$-\, e_i^* = \psi\left(\frac{-e_i}{\sigma^{(m)}}\right) \sigma^{(m)} . \tag{364.8}$$

Hence, for simultaneously computing the robust estimates for the parameter vector $\boldsymbol{\beta}$ and the standard deviation σ we apply (364.5), (363.7), (364.8), (363.9), (363.10) and (363.11).

365 Predictive Distributions for Data Points

As already mentioned in Section 361, robust estimation is generally applied to identify outliers or to avoid distorting the parameter estimation by outliers. After an observation has been marked as a possible outlier by the Winsorization according to (363.8) or (364.8), the question arises whether to reject the observation or to leave it in the data set for an ensuing parameter estimation, for instance, by the method of least squares.

Such a question may be answered by means of the predictive distributions for the observations. We want to predict actual observations. According to (314.5), we therefore compute from (262.2) the predictive distributions of the observations, which were identified as possible outliers by the Winsorization. A confidence interval for the predicted observation is then computed by means of the predictive distribution. If the actual observation lies outside the confidence interval, it is considered an outlier.

Let the observation which is identified as a possible outlier be denoted by y_k. Let y_{ku} be the observation predicted for y_k. For y_{ku}, we assume the same distribution as for y_k, so that the density $p(y_{ku}|\boldsymbol{\beta},\sigma)$ of the predicted observation y_{ku} follows from (362.2) by

$$p(y_{ku}|\boldsymbol{\beta},\sigma) \propto \frac{1}{\sigma}\, g\left(\frac{y_{ku}-f_k(\boldsymbol{\beta})}{\sigma}\right). \tag{365.1}$$

We suppose the standard deviation σ as being known, the posterior density function $p(\boldsymbol{\beta}|\mathbf{y})$ for the parameter vector $\boldsymbol{\beta}$ then follows from (363.2). Thus, we obtain the predictive density function for the unobserved data point y_{ku} with (262.2) by

$$p(y_{ku}|\mathbf{y}) \propto \int_{-\infty}^{\infty} \cdots \int_{-\infty}^{\infty} g\left(\frac{y_{ku}-f_k(\boldsymbol{\beta})}{\sigma}\right) \prod_{i=1}^{n} g\left(\frac{y_i-f_i(\boldsymbol{\beta})}{\sigma}\right) d\beta_1 \ldots d\beta_u . \tag{365.2}$$

By means of this distribution we may estimate y_{ku} or establish a confidence interval for y_{ku}. If this interval contains the observed data point y_k, we decide that y_k is not an outlier. If y_k lies outside the interval, we consider it an outlier. The distribution (365.2) is analytically not tractable, the numerical techniques of Section 27 therefore have to be applied to estimate y_{ku}, to establish confidence intervals or to test hypotheses.

The results obtained by the predictive distribution (365.2) of course depend on the constants in (362.5) or (362.6) of the distribution chosen for the data.

Example 1: The abscissae x_i for six points are given and the ordinates y_i are independently measured with the given standard deviation of $\sigma=0.02$. The observations are shown in the following table. A straight line shall be fitted through the data. Inspecting the data

x_i	1	2	3	4	5	6
y_i	0.48	1.15	1.49	2.02	2.51	3.01

reveals what seems to be an outlier for the observation y_2. Thus, we will apply a robust estimation and compute a confidence interval for the predicted observation y_{2u}, if the Winsorization marks y_2 as a possible outlier.

The observation equations for fitting a straight line follow by

$$y_i + e_i = \beta_o + x_i\beta_1 \quad \text{with} \quad i \in \{1,\ldots,6\},$$

where β_o and β_1 denote the unknown parameters.

With

$$\mathbf{y} + \mathbf{e} = \mathbf{X}\boldsymbol{\beta}$$

the least squares fit is obtained from (312.11) with

$$\hat{\boldsymbol{\beta}} = (\mathbf{X}'\mathbf{X})^{-1}\mathbf{X}'\mathbf{y}$$

and

$$\hat{\beta}_o = 0.051, \quad \hat{\beta}_1 = 0.493. \tag{365.3}$$

We compute Huber's robust M-estimate with c=1.5 and use $\beta_o^{(1)}=\hat{\beta}_o$, $\beta_1^{(1)}=\hat{\beta}_1$ as first approximation. We iteratively apply (363.7)

$$-e_i = y_i - \beta_o^{(m)} - x_i\beta_1^{(m)}$$

and (363.8) together with (363.5)

$$\begin{aligned} -e_i^* = \psi\left(\frac{-e_i}{\sigma}\right)\sigma &= -c\sigma \quad \text{for} \quad -e_i < -c\sigma \\ &= -e_i \quad \text{for} \quad |e_i| \le c\sigma \\ &= c\sigma \quad \text{for} \quad -e_i > c\sigma. \end{aligned}$$

The normal equations (363.10)

$$\mathbf{X}'\mathbf{X}\boldsymbol{\tau} = \mathbf{X}'\mathbf{r}^*$$

are solved, to give

$$\boldsymbol{\beta}^{(m+1)} = \boldsymbol{\beta}^{(m)} + \boldsymbol{\tau}.$$

After the change of the parameters is less than ε times their standard deviations

$$|\tau_i| < \varepsilon\sigma(\bar{x}_{ii})^{1/2},$$

where $\boldsymbol{\tau}=(\tau_i)$ and $(\mathbf{X}'\mathbf{X})^{-1}=(\bar{x}_{ij})$, the iterations are stopped. With ε=0.0001 twelve iterations are needed to obtain the robust estimates

$$\beta_o^{(12)} = -0.004, \quad \beta_1^{(12)} = 0.503. \tag{365.4}$$

Through all iterations the residual for the observation y_2 was changed by the Winsorization.

The predictive density function for the unobserved data point y_{2u} follows from (365.2) with (362.3) and (362.5) by

$$p(y_{2u}|\mathbf{y}) \propto \int_{-\infty}^{\infty} \int_{\infty}^{\infty} \exp[-\frac{1}{2}((y_{2u}-\beta_o-x_2\beta_1)/\sigma)^2]$$

$$\exp[-\frac{1}{2} \sum_{i=1}^{6} ((y_i-\beta_o-x_i\beta_1)/\sigma)^2] d\beta_1 d\beta_2$$

$$\text{for} \quad |y_k-\beta_o-x_k\beta_1| \le c\sigma$$

$$\propto \int_{-\infty}^{\infty} \int_{-\infty}^{\infty} \exp[-\frac{1}{2}(2c|(y_{2u}-\beta_o-x_2\beta_1)/\sigma|-c^2)]$$

$$\exp[-\frac{1}{2} \sum_{i=1}^{6} (2c|(y_i-\beta_o-x_i\beta_1)/\sigma|-c^2)] d\beta_1 d\beta_2$$

$$\text{for} \quad |y_k-\beta_o-x_k\beta_1| > c\sigma.$$

By means of this density function we will numerically compute the Bayes estimate $\hat{y}_{2uB}$ of y_{2u} by (272.3) and the confidence interval by (272.6) and (272.7). The density function $p(y_{2u}|\mathbf{y})$, which is obtained as a marginal density, is computed by (273.2). We put $i=200$ and $j=200\times200$ in (273.2), which means, we generate 200 random numbers with uniform distributions for y_{2u}, 200 random numbers for β_o and for each of these random numbers 200 random numbers with uniform distributions for β_1.

Since the marginal density function of y_{2u} is univariate, the random numbers generated for y_{2u} were ordered according to increasing values and along with them the density values. Ten additional density values were then linearly interpolated between all consecutive pairs of density values. The confidence limits were obtained from (272.6) together with (272.7) by starting from the densities for the minimum and maximum values generated for y_{2u} and by checking numerically whether the inequality in (241.2) was fulfilled.

The intervals in (271.4) on the coordinate axes for β_o and β_1, which define the domain of the integration, were determined in analogy to (272.1) by finding the density values, which cease to contribute to the integrations. The interval for y_{2u} was set up such that the interval of the integration is ten to twenty per cent wider than the confidence interval for y_{2u}, which ensures that only those density values are neglected which do not contribute to the integration.

For each of the following intervals defining the domain of the integration

$$0.95 < y_{2u} < 1.05, \ -0.026 < \beta_o < 0.030, \ 0.494 < \beta_1 < 0.509$$

and

$$0.96 < y_{2u} < 1.04, \ -0.023 < \beta_o < 0.027, \ 0.495 < \beta_1 < 0.508$$

two sets of 200×200×200 random numbers were generated and $\hat{y}_{2uB}$ and the confidence limits for y_{2u} with content 0.95 were computed. The results are given in the table (365.5). Although relatively few random numbers were generated for y_{2u}, β_o and β_1, the results in (365.5) between the different sets of random numbers agree very well. As could be expected, $\hat{y}_{2uB}$ comes close to $\beta_o^{(12)}+2\beta_1^{(12)}=1.002$.

	$\hat{y}_{2uB}$	Confidence Interval
	0.999	$0.967 < y_{2u} < 1.030$
	0.999	$0.969 < y_{2u} < 1.030$
	0.998	$0.968 < y_{2u} < 1.030$
	1.001	$0.969 < y_{2u} < 1.028$
Mean	0.999	$0.968 < y_{2u} < 1.030$

(365.5)

Furthermore, the marginal posterior density function $p(y_{2u}|y)$ of y_{2u} was approximately computed with (274.6) by introducing for β_o and β_1 the maximum likelihood estimate, that is the robust estimate (365.4). For each of the two intervals defining the domain of the integration for y_{2u}

$$0.95 < y_{2u} < 1.05, \quad 0.96 < y_{2u} < 1.04$$

two sets of 500 random numbers for y_{2u} were generated giving the results (365.6). The values in (365.5) and (365.6) for the Bayes estimate $\hat{y}_{2uB}$ and the confidence interval of

	$\hat{y}_{2uB}$	Confidence Interval
	1.002	$0.975 < y_{2u} < 1.030$
	1.002	$0.973 < y_{2u} < 1.032$
	1.001	$0.975 < y_{2u} < 1.029$
	1.002	$0.975 < y_{2u} < 1.029$
Mean	1.002	$0.974 < y_{2u} < 1.030$

(365.6)

y_{2u} agree very well. The observations therefore contain enough information on the unknown parameters β_o and β_1 to warrant an approximate computation of the marginal distribution.

The confidence interval computed for the predicted observation y_{2u} does not contain the value of the observation y_2, which is $y_2=1.15$. We therefore may consider this observation an outlier.

As already mentioned, the confidence interval computed for any predicted observation of this example depends on the constant c in (362.5). In fact, the length of the confidence interval computed for y_{2u} is approximately equal to $2c\sigma=0.06$. In other words, if the residual for the observation y_2 is changed by the Winsorization, y_2 is assumed to be an outlier. Δ

37 Reconstruction of Digital Images

371 Model Description and Likelihood Functions

The processing of digital images has been intensively investigated in many scientific disciplines for a great variety of applications. For instance, photogrammetry with its geodetic applications has participated in the new developments and is actively engaged to try new developments in the field of image processing (Förstner 1988).

Numerous methods exist for the analysis of digital images based on different assumptions. While there is a similarity between the models which define the mathematical relation between the data and the unknown parameters, the statistical assumptions concerning the data and the methods for estimating the unknown parameters differ vastly. The Bayesian approach, due to its flexibility of introducing prior information for the parameters, unites some of the approaches and puts them on a sound theoretical basis. It is therefore presented here.

An object radiates energy, and if this energy passes an image recording system, an image is formed. A digital image consists of a rectangular array of picture elements, the so-called *pixels*. For each pixel, due to the different intensity of the energy radiated by the object, different gray levels are measured and digitally recorded.

The imaging systems do not work perfectly. We therefore have to differentiate between the measured gray levels for the pixels, which shall be collected in the vector $\mathbf{y}$ of observations, and the unknown parameters representing the gray levels obtained under ideal conditions. These unknown parameters or signals, collected in $\boldsymbol{\beta}$, have to be estimated.

We will assume that given nonlinear functions $\mathbf{f}(\boldsymbol{\beta})$ with $\mathbf{f}(\boldsymbol{\beta})=(f_i(\boldsymbol{\beta}))$ of the unknown parameters $\boldsymbol{\beta}$ represent the imaging system and give the mathematical relation between the unknown signal $\boldsymbol{\beta}$ and the observations $\mathbf{y}$. The functions $\mathbf{f}(\boldsymbol{\beta})$ therefore account for the distortions or the blurring of the image. In addition, let the covariance matrix of the observations be defined by the unknown variance factor σ^2 and the known weight matrix $\mathbf{P}$. Thus, we obtain the nonlinear model, already introduced by (321.1),

$$E(\mathbf{y}|\boldsymbol{\beta}) = \mathbf{f}(\boldsymbol{\beta}) \quad \text{with} \quad D(\mathbf{y}|\sigma^2) = \sigma^2\mathbf{P}^{-1}. \tag{371.1}$$

Gray levels are generally restricted to values from 0 to 255. The vectors $\mathbf{y}$ and $\boldsymbol{\beta}$ may therefore be interpreted as discrete random vectors.

In many cases the nonlinear function $\mathbf{f}(\boldsymbol{\beta})$ can be approximated by the linear function $\mathbf{X}\boldsymbol{\beta}$, for instance, if the blurring of an image can be modeled by a shift-invariant point-spread function. In such a case the matrix $\mathbf{X}$ might achieve a simple convolution over a small window, for instance a weighted mean of a pixel with its eight nearest neighbors. In the simplest case $\mathbf{X}$ is identical with the identity matrix $\mathbf{I}$

$$\mathbf{X} = \mathbf{I}. \tag{371.2}$$

By means of the linear functions $\mathbf{X}\boldsymbol{\beta}$ we obtain the linear model, see also (311.3),

$$E(\mathbf{y}|\boldsymbol{\beta}) = \mathbf{X}\boldsymbol{\beta} \quad \text{with} \quad D(\mathbf{y}|\sigma^2) = \sigma^2\mathbf{P}^{-1}. \tag{371.3}$$

We will also consider models where the variance factor σ^2 is known, so that the covariance matrix $\boldsymbol{\Sigma}$ of the observations $\mathbf{y}$ is given by

$$\boldsymbol{\Sigma} = \sigma^2\mathbf{P}^{-1}. \tag{371.4}$$

This leads instead of (371.1) and (371.3) to the models

$$E(\mathbf{y}|\boldsymbol{\beta}) = \mathbf{f}(\boldsymbol{\beta}) \quad \text{with} \quad D(\mathbf{y}) = \boldsymbol{\Sigma} \tag{371.5}$$

and

$$E(\mathbf{y}|\boldsymbol{\beta}) = \mathbf{X}\boldsymbol{\beta} \quad \text{with} \quad D(\mathbf{y}) = \boldsymbol{\Sigma}. \tag{371.6}$$

We will also introduce the very simple covariance matrix

$$\boldsymbol{\Sigma} = \mathbf{I}. \tag{371.7}$$

The observations $\mathbf{y}$ are assumed as normally distributed, so that the likelihood function follows with (A21.1) for the nonlinear model (371.5) by

$$p(\mathbf{y}|\boldsymbol{\beta}) \propto \exp[-\frac{1}{2}(\mathbf{y}-\mathbf{f}(\boldsymbol{\beta}))'\boldsymbol{\Sigma}^{-1}(\mathbf{y}-\mathbf{f}(\boldsymbol{\beta}))] \tag{371.8}$$

and for the linear model (371.6) by

$$p(\mathbf{y}|\boldsymbol{\beta}) \propto \exp[-\frac{1}{2}(\mathbf{y}-\mathbf{X}\boldsymbol{\beta})'\boldsymbol{\Sigma}^{-1}(\mathbf{y}-\mathbf{X}\boldsymbol{\beta})]. \tag{371.9}$$

The likelihood functions for the models (371.1) and (371.3) are correspondingly obtained.

Depending on the choice of the prior distribution $p(\boldsymbol{\beta})$, Bayes' theorem (211.1) leads to different posteriori distributions $p(\boldsymbol{\beta}|\mathbf{y})$ of the signal $\boldsymbol{\beta}$ and thus to different methods of the image restoration. In the following we will discuss three different choices of prior distributions. The first one agrees with the priors of the linear model, the second one is based on the normal distribution and the Gibbs distribution, the third choice leads to the maximum entropy method.

372 Normal-Gamma Distribution as Prior

We will assume that the linear model (371.3) describes the imaging system. If prior information on the unknown signal $\boldsymbol{\beta}$ is available by the expected value $\boldsymbol{\mu}_p$ and the covariance matrix $\boldsymbol{\Sigma}_\beta$

$$E(\boldsymbol{\beta}) = \boldsymbol{\mu}_p, \quad D(\boldsymbol{\beta}) = \boldsymbol{\Sigma}_\beta, \tag{372.1}$$

and if also the expected value and the variance of the variance factor σ^2 are known in advance, we may introduce the normal-gamma distribution (313.1) as prior distribution, whose parameters are determined by (313.2). The Bayes estimate $\hat{\boldsymbol{\beta}}_B$ of the unknown parameters $\boldsymbol{\beta}$ then follows from (313.7), where we omit the bars in agreement with the notation of (371.3),

$$\hat{\boldsymbol{\beta}}_B = (\mathbf{X}'\mathbf{P}\mathbf{X}+\mathbf{V}^{-1})^{-1}(\mathbf{X}'\mathbf{P}\mathbf{y}+\mathbf{V}^{-1}\boldsymbol{\mu}). \tag{372.2}$$

For the special case $\mathbf{X}=\mathbf{I}$ from (371.2) we obtain

$$\hat{\boldsymbol{\beta}}_B = (\mathbf{P}+\mathbf{V}^{-1})^{-1}(\mathbf{P}\mathbf{y}+\mathbf{V}^{-1}\boldsymbol{\mu}) \tag{372.3}$$

and recognize the Bayes estimate $\hat{\boldsymbol{\beta}}_B$ being the weighted mean of the prior information $\boldsymbol{\mu}=\boldsymbol{\mu}_p$ and of the data $\mathbf{y}$, where $\mathbf{P}$ and $\mathbf{V}^{-1}$ are the weight matrices.

Reconstructing images by means of (372.2) or (372.3) is not very efficient, since according to (372.1) the prior information on the gray levels and their variances and covariances has to be introduced for each individual pixel. This is time-consuming and the approach is therefore not well suited for an automatic processing of images. It has to be restricted to special cases.

373 Normal Distribution and Gibbs Distribution as Priors

We start from the simplified nonlinear model (371.5) and introduce the prior information on the unknown parameters $\boldsymbol{\beta}$ as in (372.1) by the expected values $\boldsymbol{\mu}_p$ and the covariance matrix $\boldsymbol{\Sigma}_\beta$. If in addition the parameters $\boldsymbol{\beta}$ are normally distributed, the prior distribution $p(\boldsymbol{\beta})$ for $\boldsymbol{\beta}$ results with (A21.1) from

$$p(\boldsymbol{\beta}) \propto \exp(-\frac{1}{2}(\boldsymbol{\beta}-\boldsymbol{\mu}_p)'\boldsymbol{\Sigma}_\beta^{-1}(\boldsymbol{\beta}-\boldsymbol{\mu}_p)) \tag{373.1}$$

or from

$$p(\boldsymbol{\beta}) \propto \exp(-U(\boldsymbol{\beta})) \tag{373.2}$$

with

$$U(\boldsymbol{\beta}) = \frac{1}{2}(\boldsymbol{\beta}-\boldsymbol{\mu}_p)'\boldsymbol{\Sigma}_\beta^{-1}(\boldsymbol{\beta}-\boldsymbol{\mu}_p), \tag{373.3}$$

where $U(\boldsymbol{\beta})$ is a scalar function of the unknown signal $\boldsymbol{\beta}$.

If our prior knowledge $\boldsymbol{\mu}_p$ agrees well with the signal $\boldsymbol{\beta}$ to be reconstructed, the function $U(\boldsymbol{\beta})$ will have a small value. In general, images are smooth, so that our prior assumption of an image is a smooth picture. The function $U(\boldsymbol{\beta})$ in (373.3) therefore measures the roughness of an image, and $U(\boldsymbol{\beta})$ is called the energy attributed to the signal $\boldsymbol{\beta}$. We will prefer images with small energies as compared to images with large energies.

If the covariance matrix $\boldsymbol{\Sigma}_\beta$ in (373.3) has a diagonal form, we may represent the energy $U(\boldsymbol{\beta})$ by

$$U(\boldsymbol{\beta}) = \sum_i U_i(\boldsymbol{\beta}), \tag{373.4}$$

where the summation is extended over each pixel i of the image and $U_i(\boldsymbol{\beta})$ denotes the contribution of the pixel i to the energy $U(\boldsymbol{\beta})$. We compute $U_i(\boldsymbol{\beta})$ locally from the differences of the gray levels of the pixels, which are neighbors of the pixel i. The function $U(\boldsymbol{\beta})$ then gives a measure of the roughness of the image. We define a neighborhood of the pixel i by the four pixels above, below and on both sides of the pixel. This is shown

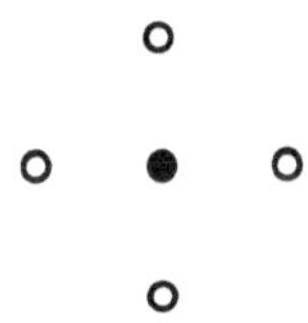

Fig. 373-1

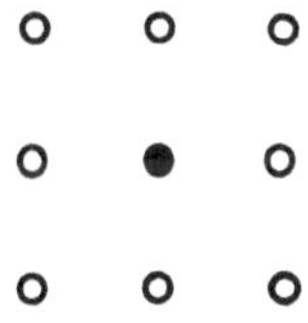

Fig. 373-2

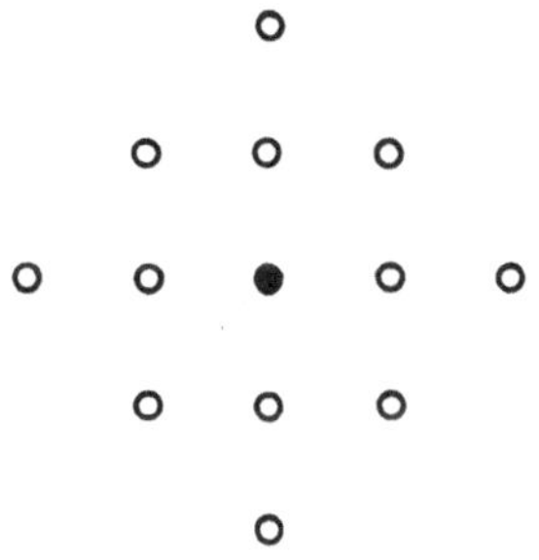

Fig. 373-3

in Fig. 373-1, where the pixel i, for which we introduce the neighborhood, is indicated by a dot and the neighbors by circles. Larger neighborhoods are depicted in Figs. 373-2 and 373-3.

The contribution $U_i(\boldsymbol{\beta})$ in (373.4) to the energy $U(\boldsymbol{\beta})$ is now obtained by summing the square of the difference $\beta_j - \beta_i$ of the gray levels between the pixel i and a pixel j over all pixels j in the neighborhood of pixel i

$$U_i(\boldsymbol{\beta}) = \sum_j (\beta_j - \beta_i)^2. \tag{373.5}$$

Large differences of gray levels give large contributions $U_i(\boldsymbol{\beta})$ and therefore high energies, which in turn lead to small values of the prior distribution $p(\boldsymbol{\beta})$ in (373.2). On the other hand, small differences of gray levels give high density values. This agrees with our prior conception that an image is smooth. Of course, different functions of $\beta_j - \beta_i$ could have been chosen in (373.5), to express the contribution $U_i(\boldsymbol{\beta})$ of the pixel i to the energy $U(\boldsymbol{\beta})$.

The local representation of the prior distribution (373.2) by (373.4) and (373.5) was given here by a heuristic argument. However, this distribution can also be derived, if we assume a Markoff random field for the unknown parameters β_i with $\boldsymbol{\beta}=(\beta_i)$ such that the probability

$$P(\beta_i \in D | \beta_j,\ i \neq j) \text{ depends only on } \beta_j \text{ with j being a neighbor of i}, \quad (373.6)$$

where D denotes a subspace of the space for β_i. The prior distribution $p(\boldsymbol{\beta})$ is then given by the Gibbs distribution (Spitzer 1971), whose functional form is identical with (373.2). The energy $U(\boldsymbol{\beta})$ is locally defined and may be computed by (373.5) or similar expressions (Geman and Geman 1984; Geman et al. 1987).

As mentioned, we assumed (371.5) as the model for the image reconstruction. If the covariance matrix $\boldsymbol{\Sigma}$ of the observations $\mathbf{y}$ has a diagonal form

$$\boldsymbol{\Sigma} = \mathrm{diag}(\sigma_1^2, \ldots, \sigma_i^2, \ldots), \quad (373.7)$$

the likelihood function $p(\mathbf{y}|\boldsymbol{\beta})$ is then given with (371.8) by

$$p(\mathbf{y}|\boldsymbol{\beta}) \propto \exp[-\sum_i (y_i - f_i(\boldsymbol{\beta}))^2/(2\sigma_i^2)]. \quad (373.8)$$

If $f_i(\boldsymbol{\beta})$ can be locally computed, which will be assumed, for instance from the pixel i and its neighbors, then the likelihood function (373.8) is given in a local representation. In the simplest case we have

$$f_i(\boldsymbol{\beta}) = \beta_i. \quad (373.9)$$

Bayes' theorem (211.1) leads from the prior density (373.2) together with (373.4), (373.5) and the likelihood function (373.8) to the posterior density function $p(\boldsymbol{\beta}|\mathbf{y})$ of the unknown signal $\boldsymbol{\beta}$ given in a local representation, if we introduce a constant b to weigh $U_i(\boldsymbol{\beta})$ with respect to the likelihood function,

$$p(\boldsymbol{\beta}|\mathbf{y}) \propto \exp\{-\sum_i [b \sum_j (\beta_j - \beta_i)^2 + (y_i - f_i(\boldsymbol{\beta}))^2/(2\sigma_i^2)]\}. \quad (373.10)$$

The first expression in the exponent of $p(\boldsymbol{\beta}|\mathbf{y})$ measures, as mentioned, the roughness or the energy of the signal $\boldsymbol{\beta}$, while the second term gives a measure for the fidelity of the observations $\mathbf{y}$ to the signal $\boldsymbol{\beta}$. Thus, the posterior distribution $p(\boldsymbol{\beta}|\mathbf{y})$ is a function of the roughness and the fidelity of the signal $\boldsymbol{\beta}$. Based on this posterior distribution, the image is reconstructed by means of the Bayes estimate $\hat{\boldsymbol{\beta}}_B$ from (231.5) or the MAP estimate $\bar{\boldsymbol{\beta}}$ from (232.4) of the unknown signal $\boldsymbol{\beta}$.

In general, the posterior distribution (373.10) is analytically not tractable, so that the estimates $\hat{\boldsymbol{\beta}}_B$ or $\bar{\boldsymbol{\beta}}$ have to be computed numerically. But even a small digital picture may have 512x512 pixels and therefore the same number of unknown parameters $\boldsymbol{\beta}$. A computation of $\hat{\boldsymbol{\beta}}_B$ from (272.3) or a numerical derivation of $\bar{\boldsymbol{\beta}}$ would involve a tremendous amount of computational work. But the numerical effort can be considerably reduced, if we take advantage of the local representation of the posterior distribution (373.10). This is done for computing the Bayes estimate $\hat{\boldsymbol{\beta}}_B$ in the following approach.

We compute the Bayes estimate of each parameter β_i by means of its marginal posterior distribution $p(\beta_i|\mathbf{y})$, which we approximately obtain from (274.7). Since the Bayes estimate of $\boldsymbol{\beta}$ is not known in advance, we have to go through iterations. Let $\hat{\boldsymbol{\beta}}_{oB}$ with $\hat{\boldsymbol{\beta}}_{oB}=(\hat{\beta}_{ioB})$ be the Bayes estimate of $\boldsymbol{\beta}$ in the oth iteration. For the next iteration o+1 the marginal density $p(\beta_i|\mathbf{y})$ is therefore obtained from (274.7) by

$$p(\beta_i|\mathbf{y}) \propto p(\hat{\beta}_{1oB},\hat{\beta}_{2oB},\ldots,\hat{\beta}_{i-1,oB},\beta_i,\hat{\beta}_{i+1,oB},\ldots|\mathbf{y}) \tag{373.11}$$

or with (373.9) and (373.10), if q denotes the number of pixels in the neighborhood of pixel i

$$\begin{aligned} p(\beta_i|\mathbf{y}) &\propto \exp\{-[b\sum_j(\hat{\beta}_{joB}-\beta_i)^2+(y_i-\beta_i)^2/(2\sigma_i^2)]\} \\ &\propto \exp\{-[(bq+1/(2\sigma_i^2))\beta_i^2-2(b\sum_j\hat{\beta}_{joB}+y_i/(2\sigma_i^2))\beta_i]\}. \end{aligned} \tag{373.12}$$

With completing the square on β_i, we obtain from (A11.1) for β_i the normal distribution

$$\beta_i|\mathbf{y} \sim N((b\sum_j\hat{\beta}_{joB}+y_i/(2\sigma_i^2))(bq+1/(2\sigma_i^2))^{-1},\ (2bq+1/\sigma_i^2)^{-1}) \tag{373.13}$$

and the Bayes estimate $\hat{\beta}_{i,o+1,B}$ for β_i of the (o+1)th iteration from (231.5) and (A11.3) by

$$\hat{\beta}_{i,o+1,B} = (b\sum_j\hat{\beta}_{joB}+y_i/(2\sigma_i^2))(bq+1/(2\sigma_i^2))^{-1}. \tag{373.14}$$

The estimate $\hat{\beta}_{i,o+1,B}$ is then substituted in (373.11) and the next pixel k is processed. The marginal density $p(\beta_k|\mathbf{y})$ of β_k for this pixel follows with

$$p(\beta_k|\mathbf{y}) \propto p(\hat{\beta}_{1oB},\hat{\beta}_{2oB},\ldots,\hat{\beta}_{i-1,oB},\hat{\beta}_{i,o+1,B},\hat{\beta}_{i+1,oB},\ldots,\beta_k,\ldots|\mathbf{y}). \tag{373.15}$$

This procedure is repeated until the last pixel and $\hat{\boldsymbol{\beta}}_{o+1,B}$ of the (o+1)th iteration is obtained. The sequence of the processing of the pixels is random to avoid any systematic effect. Since gray levels are estimated, the value for $\hat{\beta}_{ioB}$ is rounded to the next integer.

At the first iteration, when estimates have not been computed yet, we use the observation y_i with $\mathbf{y}=(y_i)$ as estimate for β_i. As soon as the estimates from one iteration to the next do not change any more, the iterations are stopped.

If functions are used different from (373.5) and (373.9) to express the contribution $U_i(\boldsymbol{\beta})$ to the enery $U(\boldsymbol{\beta})$ and to model the imaging system, it may not be possible any more to derive the expected value of the distribution $p(\beta_i|\mathbf{y})$ analytically as in (373.14). Then the estimates $\hat{\beta}_{ioB}$ have to be numerically computed from (272.3) with (373.11) and (373.15).

A similar approach to compute the MAP estimate $\bar{\boldsymbol{\beta}}$ of the signal $\boldsymbol{\beta}$ is proposed in Geman and Geman (1984). Because of its analogy to a process in chemistry to achieve a state of low energy by heating and slowly cooling a substance, it is called simulated annealing (Aarts and van Laarhoven 1987; Ripley 1988, p.95). The observations y_i are introduced as approximate values for the estimates of the parameters β_i. The pixels are then visited randomly and a random value is chosen from the distribution (373.10) for β_i. With each iteration, which involves the visit of each pixel, the temperature T introduced into the exponent of (373.10) by

$$-\sum_i [b \sum_j (\beta_j-\beta_i)^2 + (y_i-f_i(\boldsymbol{\beta}))^2/(2\sigma_i^2)]/T$$

is decreased, until a state of low energy and therefore the maximum of the posterior density is attained. This procedure requires more iterations and therefore more computing time than the approach based on (373.15), although the results are similar (Busch and Koch 1990).

Images very often contain edges, which have to be considered in a complete reconstruction of an image, since the smoothing of a picture may not be extended across the edges. To solve this problem, we have to introduce line elements as unknown parameters in addition to the unknown gray levels of the pixels considered so far. These unknown line elements take the value one, if they are present, or zero, if they are absent. The line elements are placed between the pixels, as shown in Fig. 373-4.

Another example where the need for additional unknown parameters arises is the discrimination of textures. Metal, plastic and wooden objects, for instance, have to be identified in an image. An additional unknown parameter is then defined for each pixel. The value of this parameter indicates to which texture the pixel belongs.

Let $\boldsymbol{\beta}_a$ denote the vector of additional parameters, which are discrete random variables. The prior distribution $p(\boldsymbol{\beta},\boldsymbol{\beta}_a)$ of the signal $\boldsymbol{\beta}$ and the additional parameters $\boldsymbol{\beta}_a$ shall be

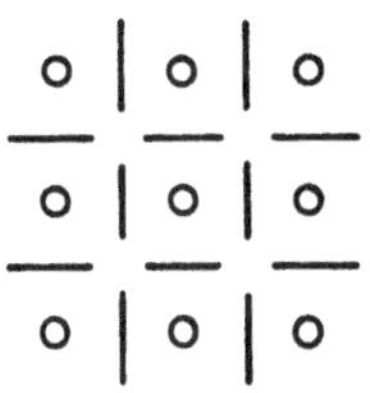

Fig. 373-4

given with (211.2) by

$$p(\boldsymbol{\beta},\boldsymbol{\beta}_a) = p(\boldsymbol{\beta}|\boldsymbol{\beta}_a)p(\boldsymbol{\beta}_a), \tag{373.16}$$

where $p(\boldsymbol{\beta}|\boldsymbol{\beta}_a)$ denotes the conditional distribution of $\boldsymbol{\beta}$ given $\boldsymbol{\beta}_a$.

If the additional parameters $\boldsymbol{\beta}_a$ consist of line elements, the density function $p(\boldsymbol{\beta}|\boldsymbol{\beta}_a)$ is still represented by (373.2) together with (373.4) and (373.5). However, differences of gray levels across a line element may not contribute in (373.5) to the energy of the signal. Thus,

$$p(\boldsymbol{\beta}|\boldsymbol{\beta}_a) \propto \exp[-\sum_i U_i(\boldsymbol{\beta}|\boldsymbol{\beta}_a)] \tag{373.17}$$

with

$$U_i(\boldsymbol{\beta}|\boldsymbol{\beta}_a) = (b/q_a)\sum_j (1-\beta_{ija})(\beta_j-\beta_i)^2, \tag{373.18}$$

where β_{ija} with $\boldsymbol{\beta}_a=[\ldots,\beta_{ija},\ldots]'$ denotes the line element between the pixel i and j with value one, if it is present, and value zero, if it is absent, and q_a the number of line elements in the neigborhood of pixel i with $\beta_{ija}=0$. The constant b is again as in (373.10) a weight factor.

The density function $p(\boldsymbol{\beta}_a)$ of the additional unknown parameters is also locally represented with (373.2) and (373.4) by

$$p(\boldsymbol{\beta}_a) \propto \exp[-\sum_k U_k(\boldsymbol{\beta}_a)] \tag{373.19}$$

with as many functions $U_k(\boldsymbol{\beta}_a)$ as unknown additional parameters $\boldsymbol{\beta}_a$. In the case of line elements we choose

$$U_k(\boldsymbol{\beta}_a) = c\sum_l f_l(\beta_{ija},\beta_{mna}), \tag{373.20}$$

where the summation is extended over special configurations of line elements β_{mna} in the neighborhood of the line element β_{ija}. The function $f_l(\beta_{ija},\beta_{mna})$ attributes low energy or large weight to lines which continue and high energy or low weight to isolated lines and to beginnings and ending of lines or unlikely configurations of lines. The constant c weighs the contribution $U_k(\boldsymbol{\beta}_a)$ to the energy with respect to $U_i(\boldsymbol{\beta}|\boldsymbol{\beta}_a)$ from (373.18). More detailed suggestions for representing the prior density function in the case of line elements or texture labels can be found in Geman et al. (1987) and in the case of line elements in Busch and Koch (1990).

The likelihood function $p(\mathbf{y}|\boldsymbol{\beta})$ of the observations $\mathbf{y}$ is not affected by the additional parameters. With

$$p(\mathbf{y}|\boldsymbol{\beta},\boldsymbol{\beta}_a) = p(\mathbf{y}|\boldsymbol{\beta}) \tag{373.21}$$

we therefore obtain from the local representation (373.8)

$$p(\mathbf{y}|\boldsymbol{\beta},\boldsymbol{\beta}_a) \propto \exp[-\sum_i (y_i-f_i(\boldsymbol{\beta}))^2/(2\sigma_i^2)]. \tag{373.22}$$

Bayes' theorem (211.1) now leads with (373.16) to (373.20) and with (373.22) to the posterior distribution of the parameters $\boldsymbol{\beta}$ and the additional line elements $\boldsymbol{\beta}_a$ given in a local representation

$$p(\boldsymbol{\beta},\boldsymbol{\beta}_a|\mathbf{y}) \propto \exp\{-\sum_i [(b/q_a)\sum_j (1-\beta_{ija})(\beta_j-\beta_i)^2 + (y_i-f_i(\boldsymbol{\beta}))^2/(2\sigma_i^2)] - \sum_k [c\sum_l f_l(\beta_{ija},\beta_{mna})]\}. \tag{373.23}$$

We compute the approximate marginal distribution $p(\beta_i|\mathbf{y})$ for the gray level β_i of pixel i correspondingly to (373.11) with $\hat{\boldsymbol{\beta}}_{aoB}=(\hat{\beta}_{ijaoB})$ being the Bayes estimate for $\boldsymbol{\beta}_a$ of the oth iteration

$$p(\beta_i|\mathbf{y}) = p(\hat{\beta}_{1oB},\hat{\beta}_{2oB},\ldots,\hat{\beta}_{i-1,oB},\beta_i,\hat{\beta}_{i+1,oB},\ldots,\hat{\beta}_{i-1,jaoB},\hat{\beta}_{ijaoB},\hat{\beta}_{i+1,jaoB},\ldots|\mathbf{y}). \tag{373.24}$$

The approximate marginal distribution $p(\beta_{ija}|\mathbf{y})$ of the line element β_{ija} follows accordingly by

$$p(\beta_{ija}|\mathbf{y}) = p(\hat{\beta}_{1oB},\hat{\beta}_{2oB},\ldots,\hat{\beta}_{i-1,oB},\hat{\beta}_{ioB},\hat{\beta}_{i+1,oB},\ldots,\hat{\beta}_{i-1,jaoB},\beta_{ija},\hat{\beta}_{i+1,jaoB},\ldots|\mathbf{y}). \tag{373.25}$$

By substituting (373.9) and (373.24) in (373.23) we find

$$p(\beta_i|\mathbf{y}) \propto \exp\{-[(b/q_a)\sum_j(1-\hat{\beta}_{ijaoB})(\hat{\beta}_{joB}-\beta_i)^2+(y_i-\beta_i)^2/(2\sigma_i^2)]\}$$
$$\propto \exp\{-[(b+1/(2\sigma_i^2))\beta_i^2-2((b/q_a)\sum_j(1-\hat{\beta}_{ijaoB})\hat{\beta}_{joB}+y_i/(2\sigma_i^2))\beta_i]\}. \tag{373.26}$$

Thus, $p(\beta_i|\mathbf{y})$ is given by the normal distribution and the Bayes estimate $\hat{\beta}_{i,o+1,B}$ for β_i of the (o+1)th iteration follows from (231.5) with (A11.3) by

$$\hat{\beta}_{i,o+1,B} = ((b/q_a)\sum_j(1-\hat{\beta}_{ijaoB})\hat{\beta}_{joB}+y_i/(2\sigma_i^2))(b+1/(2\sigma_i^2))^{-1}. \qquad (373.27)$$

The Bayes estimate $\hat{\beta}_{ija,o+1,B}$ for the line element β_{ija} of the (o+1)th iteration is either equal to zero or equal to one and is obtained by determining the maximum of

$$p(\beta_{ija}=0|\mathbf{y}) \quad \text{and} \quad p(\beta_{ija}=1|\mathbf{y}).$$

In a random sequence the Bayes estimates of the gray levels of all pixels and of all line elements are computed and substituted in (373.24) and (373.25). In a new iteration, again with a random sequence, the estimates for all pixels and line elements are recomputed. The iterations stop, when no changes in the estimates occur between two iterations. At the beginning of the iterations it is assumed that no line elements are present. Examples for this method of image restoration can be found in Busch and Koch (1990).

374 Prior Leading to Maximum Entropy Restoration

The representation (373.2) of the prior distribution by means of the roughness of an image suggests introducing the entropy as a measure of roughness. If we interpret the gray level values β_i with $\boldsymbol{\beta}=(\beta_i)$ as discrete densities, the information H_n or the uncertainty of an image follows from (223.3) by

$$H_n = -\sum_i \beta_i \ln\beta_i, \qquad (374.1)$$

where the summation is extended over each pixel i. By defining a prior density function $p(\boldsymbol{\beta})$ with

$$p(\boldsymbol{\beta}) \propto \exp(-\lambda \sum_i \beta_i \ln\beta_i) \quad \text{with} \quad \lambda > 0, \qquad (374.2)$$

where λ is a constant, we use large prior density values for large values of the entropy or a large amount of uncertainty. In a reconstruction we therefore prefer images with large uncertainty. According to (223.5) these are pictures which are smooth and as uniformly gray as possible.

Using the prior density (374.2) and the likelihood function (371.9) in Bayes' theorem (211.1) leads to the posterior density function $p(\boldsymbol{\beta}|\mathbf{y})$ for the unknown signal $\boldsymbol{\beta}$

$$p(\boldsymbol{\beta}|\mathbf{y}) \propto \exp[-\lambda \sum_i \beta_i \ln\beta_i - \frac{1}{2}(\mathbf{y}-\mathbf{X}\boldsymbol{\beta})'\boldsymbol{\Sigma}^{-1}(\mathbf{y}-\mathbf{X}\boldsymbol{\beta})]. \qquad (374.3)$$

This posterior distribution can now be used to derive the Bayes estimate $\hat{\boldsymbol{\beta}}_B$ from (231.5) or the MAP estimate $\bar{\boldsymbol{\beta}}$ from (232.4) of the unknown parameters $\boldsymbol{\beta}$. For obtaining the MAP estimate $\bar{\boldsymbol{\beta}}$ we take the natural logarithm of (374.3) and set its derivative with respect to $\boldsymbol{\beta}$ equal to zero corresponding to the derivation of (363.3). We obtain

$$\partial \ln p(\boldsymbol{\beta}|\mathbf{y})/\partial\boldsymbol{\beta} = -\lambda(\mathbf{ln}\boldsymbol{\beta} + \mathbf{1}) - \mathbf{X}'\boldsymbol{\Sigma}^{-1}(\mathbf{X}\boldsymbol{\beta}-\mathbf{y}) = \mathbf{0} \qquad (374.4)$$

with

$$\mathbf{ln}\boldsymbol{\beta} = (\ln\beta_i) \quad \text{and} \quad \mathbf{1} = (1,1,\dots,1)'.$$

Furthermore, we find

$$\lambda\mathbf{ln}\boldsymbol{\beta} = -\lambda\mathbf{1} + \mathbf{X}'\boldsymbol{\Sigma}^{-1}(\mathbf{y}-\mathbf{X}\boldsymbol{\beta}),$$

so that the MAP estimate $\bar{\boldsymbol{\beta}}$ follows with

$$\bar{\boldsymbol{\beta}} = \mathbf{exp}(-1+\mathbf{X}'\boldsymbol{\Sigma}^{-1}(\mathbf{y}-\mathbf{X}\bar{\boldsymbol{\beta}})/\lambda), \qquad (374.5)$$

where $\mathbf{exp}(\dots)$ has to be interpreted in analogy to $\mathbf{ln}\boldsymbol{\beta}$.

This estimate is also obtained by the maximum entropy method for restoring images. To derive this method, the entropy H_n following from (374.1)

$$H_n = -\sum_i \beta_i \ln\beta_i$$

is maximized subject to the constraint

$$(\mathbf{y}-\mathbf{X}\boldsymbol{\beta})'\boldsymbol{\Sigma}^{-1}(\mathbf{y}-\mathbf{X}\boldsymbol{\beta}) = \chi^2_{\alpha;n}, \qquad (374.6)$$

where $\chi^2_{\alpha;n}$ denotes the lower α-percentage point of the χ^2-distribution with n degrees of freedom (Koch 1988a, p.146). This constraint introduces the fidelity of the observations mentioned in connection with (373.10). It results from the fact that, given the parameters $\boldsymbol{\beta}$, the quadratic form on the left-hand side of (374.6) has the χ^2-distribution (Koch 1988a, p.145).

To determine the extreme value of the entropy H_n, we introduce the Lagrange function $w(\boldsymbol{\beta},\lambda)$ with

$$w(\boldsymbol{\beta},\lambda) = -\sum_i \beta_i \ln\beta_i - \frac{1}{2\lambda}((\mathbf{y}-\mathbf{X}\boldsymbol{\beta})'\boldsymbol{\Sigma}^{-1}(\mathbf{y}-\mathbf{X}\boldsymbol{\beta})-\chi^2_{\alpha;n}), \qquad (374.7)$$

where $-1/2\lambda$ denotes the Lagrange multiplier. Setting the derivative $\partial w(\boldsymbol{\beta},\lambda)/\partial\boldsymbol{\beta}$ equal to zero leads to (374.4) and therefore to the estimate $\bar{\boldsymbol{\beta}}$ which is identical with the MAP estimate (374.5).

Maximization of the entropy H_n subject to the fidelity constraint (374.6) gives images with the largest amount of uncertainty compatible with the data. These images will therefore show no features for which there is no clear evidence in the data (Gull and

Skilling 1985). This certainly is an attractive property for the image reconstruction. Very convincing results of the maximum entropy restoration have been obtained in radio astronomical interferometry (Skilling and Gull 1985). As shown, the maximum entropy restoration is also obtained with the Bayesian approach, if (374.2) is chosen as prior distribution.

The computation of the MAP estimate $\boldsymbol{\beta}$ or the maximum entropy estimate by (374.5) has not been discussed yet. Since $\boldsymbol{\beta}$ appears on both sides of the equation, an iterative procedure seems advisable. However, the exponential function may introduce instabilities into the iterative procedure so that smoothing needs to be applied. A different approach maximizes the entropy H_n subject to (374.6) numerically (Skilling and Gull 1985).

A Appendix

In the Appendix the propertiers of several univariate and multivariate distributions are collected, which were referred to at various places in the text before. As has been practiced already in Section 211, we will not distinguish in our notation a random variable from the values it takes on, but use the same letter for both quantities. We start with the univariate distributions and then continue with the multivariate distributions.

A1 Univariate Distributions

A11 Univariate Normal Distribution

Definition: The random variable x is said to be *normally distributed* with the parameters μ and σ^2, which is written $x \sim N(\mu, \sigma^2)$, if its density function $p(x|\mu,\sigma^2)$ is given by

$$p(x|\mu,\sigma^2) = \frac{1}{\sqrt{2\pi}\,\sigma} e^{-(x-\mu)^2/2\sigma^2} \quad \text{for} \quad -\infty < x < \infty. \tag{A11.1}$$

It is obvious that the first condition in (211.6) is fulfilled by the density function in (A11.1). This holds true also for the second one with (Koch 1988a, p.125)

$$\frac{1}{\sqrt{2\pi}\,\sigma} \int_{-\infty}^{\infty} e^{-(x-\mu)^2/2\sigma^2} dx = 1. \tag{A11.2}$$

If $x \sim N(\mu, \sigma^2)$, then (Koch 1988a, p.138)

$$E(x) = \mu \quad \text{and} \quad V(x) = \sigma^2. \tag{A11.3}$$

The density function of a normally distributed random variable is therefore uniquely determined by its expected value and its variance.

A12 Gamma Distribution

Definition: The random variable x has the *gamma distribution* $G(b,p)$ with the real-valued parameters b and p, thus $x \sim G(b,p)$, if its density function is given by

$$p(x|b,p) = b^p x^{p-1} e^{-bx}/\Gamma(p) \quad \text{for} \quad b > 0,\ p > 0,\ 0 < x < \infty$$

and $p(x|b,p)=0$ for the remaining values of x. (A12.1)

It is obvious that $p(x|b,p) \geq 0$, and by the definition of the gamma function $\Gamma(p)$ it can be shown (Koch 1988a, p.130) that

$$\int_0^{\infty} b^p (\Gamma(p))^{-1} x^{p-1} e^{-bx} dx = 1. \tag{A12.2}$$

Hence, (211.6) is fulfilled.

With $b=1/2$ and $p=n/2$ we find the distribution which is known as *χ^2-distribution* (Koch 1988a, p.144). We write $x \sim \chi^2(n)$, if

$$p(x|n) = \frac{1}{2^{n/2}\Gamma(n/2)} x^{n/2-1} e^{-x/2} \quad \text{for} \quad 0 < x < \infty \tag{A12.3}$$

and $p(x|n)=0$ for the remaining values of x.

If the random variable x has the gamma distribution $x \sim G(b,p)$, the moment generating function $M_x(t)$ of x is given by (Koch 1988a, p.131)

$$M_x(t) = (1-t/b)^{-p} \quad \text{for} \quad t < b. \tag{A12.4}$$

The first moment $E(x)$ and the second moment $E(x^2)$ of x follow with (Koch 1988a, p.123)

$$E(x) = \left.\frac{\partial M_x(t)}{\partial t}\right|_{t=0} \quad \text{and} \quad E(x^2) = \left.\frac{\partial^2 M_x(t)}{\partial t^2}\right|_{t=0}$$

and the variance $V(x)$ of x with

$$V(x) = E[(x-E(x))^2] = E[x^2-2xE(x)+(E(x))^2] = E(x^2)-(E(x))^2. \tag{A12.5}$$

Hence,

$$\frac{\partial M_x(t)}{\partial t} = \frac{p}{b}\left(1-\frac{t}{b}\right)^{-p-1}, \quad \left.\frac{\partial M_x(t)}{\partial t}\right|_{t=0} = \frac{p}{b}$$

$$\frac{\partial^2 M_x(t)}{\partial t^2} = \frac{p}{b^2}(p+1)\left(1-\frac{t}{b}\right)^{-p-2}, \quad \left.\frac{\partial^2 M_x(t)}{\partial t^2}\right|_{t=0} = \frac{p}{b^2}(p+1)$$

so that we obtain the expected value and the variance of the random variable x having the gamma distribution $x \sim G(b,p)$

$$E(x) = p/b \quad \text{and} \quad V(x) = p/b^2. \tag{A12.6}$$

A13 Inverted Gamma Distribution

Theorem: If the random variable x has the gamma distribution $x \sim G(b,p)$, then the random variable z with $z=1/x$ has the *inverted gamma distribution*, $z \sim IG(b,p)$, with the density function

$$p(z|b,p) = \frac{b^p}{\Gamma(p)} \left(\frac{1}{z}\right)^{p+1} e^{-b/z} \quad \text{for} \quad b > 0,\ p > 0,\ 0 < z < \infty$$

and $p(z|b,p)=0$ for the remaining values of z. (A13.1)

Proof: With the transformation $x=1/z$ and its Jacobian $\det J=-1/z^2$ (Koch 1988a, p.108) we obtain instead of the density function $p(x|b,p)$ of (A12.1) the density function $p(z|b,p)$ of (A13.1). □

The expected value and the variance of a random variable z having the inverted gamma distribution $z \sim IG(b,p)$ are given by

$$E(z) = b/(p-1) \quad \text{for} \quad p > 1$$

and

$$V(z) = b^2/((p-1)^2(p-2)) \quad \text{for} \quad p > 2. \tag{A13.2}$$

The first result follows with $\Gamma(p)=(p-1)\Gamma(p-1)$ and

$$E(z) = \int_0^\infty z \frac{b^p}{\Gamma(p)} \left(\frac{1}{z}\right)^{p+1} e^{-b/z} dz = \frac{b}{p-1} \int_0^\infty \frac{b^{p-1}}{\Gamma(p-1)} \left(\frac{1}{z}\right)^p e^{-b/z} dz$$

$$= \frac{b}{p-1} \quad \text{for} \quad p > 1,$$

since the integrant represents the density function of a random variable z with $z \sim IG(b,p-1)$. The second result follows similarly with

$$E(z^2) = \int_0^\infty z^2 \frac{b^p}{\Gamma(p)} \left(\frac{1}{z}\right)^{p+1} e^{-b/z} dz$$

$$= \frac{b^2}{(p-1)(p-2)} \int_0^\infty \frac{b^{p-2}}{\Gamma(p-2)} \left(\frac{1}{z}\right)^{p-1} e^{-b/z} dz \quad \text{for} \quad p > 2$$

and with (A12.5)

$$V(z) = \frac{b^2}{(p-1)(p-2)} - \frac{b^2}{(p-1)^2} \quad \text{for} \quad p > 2.$$

A2 Multivariate Distributions

A21 Multivariate Normal Distribution

Definition: The n×1 random vector $\mathbf{x}=[x_1,\ldots,x_n]'$ is said to have a *multivariate normal distribution* $N(\boldsymbol{\mu},\boldsymbol{\Sigma})$ with the n×1 vector $\boldsymbol{\mu}$ and the n×n positive definite matrix $\boldsymbol{\Sigma}$ as parameters, thus $\mathbf{x}\sim N(\boldsymbol{\mu},\boldsymbol{\Sigma})$, if the density function $p(\mathbf{x}|\boldsymbol{\mu},\boldsymbol{\Sigma})$ of $\mathbf{x}$ is given by

$$p(\mathbf{x}|\boldsymbol{\mu},\boldsymbol{\Sigma}) = (2\pi)^{-n/2}(\det\boldsymbol{\Sigma})^{-1/2}\exp[-\tfrac{1}{2}(\mathbf{x}-\boldsymbol{\mu})'\boldsymbol{\Sigma}^{-1}(\mathbf{x}-\boldsymbol{\mu})]. \qquad (A21.1)$$

Since $\boldsymbol{\Sigma}$ is assumed to be positive definite, $\det\boldsymbol{\Sigma}>0$ and $p(\mathbf{x}|\boldsymbol{\mu},\boldsymbol{\Sigma})\geq 0$ follow. In addition we have (Koch 1988a, p.136)

$$\int_{-\infty}^{\infty}\ldots\int_{-\infty}^{\infty}\exp[-\tfrac{1}{2}(\mathbf{x}-\boldsymbol{\mu})'\boldsymbol{\Sigma}^{-1}(\mathbf{x}-\boldsymbol{\mu})]dx_1\ldots dx_n = (2\pi)^{n/2}(\det\boldsymbol{\Sigma})^{1/2}, \qquad (A21.2)$$

so that (211.6) is fulfilled.

The density function of a normally distributed random vector is uniquely determined by its vector of expected values and its covariance matrix. This is due to the following theorem (Koch 1988a, p.138).

Theorem: Let the random vector $\mathbf{x}$ be distributed according to $\mathbf{x}\sim N(\boldsymbol{\mu},\boldsymbol{\Sigma})$, then $E(\mathbf{x})=\boldsymbol{\mu}$ and $D(\mathbf{x})=\boldsymbol{\Sigma}$. (A21.3)

A22 Multivariate t-Distribution

Theorem: Let the k×1 random vector $\mathbf{z}=[z_1,\ldots,z_k]'$ be distributed according to $\mathbf{z}\sim N(\mathbf{0},\mathbf{N}^{-1})$ with the k×k matrix $\mathbf{N}$ being positive definite. Furthermore let the random variable h be distributed according to $h\sim\chi^2(\nu)$ with ν degrees of freedom. Let the random vector $\mathbf{z}$ and the random variable h be independent. Then the k×1 random vector $\mathbf{x}=[x_1,\ldots,x_k]'$, which originates from the transformation

$$x_i = z_i(h/\nu)^{-1/2} + \mu_i \quad \text{for} \quad i\in\{1,\ldots,k\},$$

has the *multivariate t-distribution* with the k×1 vector $\boldsymbol{\mu}$ from $\boldsymbol{\mu}=(\mu_i)$, the matrix $\mathbf{N}^{-1}$ and ν as parameters, abbreviated by $\mathbf{x}\sim t(\boldsymbol{\mu},\mathbf{N}^{-1},\nu)$, if the density function $p(\mathbf{x}|\boldsymbol{\mu},\mathbf{N}^{-1},\nu)$ of $\mathbf{x}$ is given by

$$p(\mathbf{x}|\boldsymbol{\mu},\mathbf{N}^{-1},\nu) = \frac{\nu^{\nu/2}\Gamma((k+\nu)/2)(\det\mathbf{N})^{1/2}}{\pi^{k/2}\Gamma(\nu/2)}(\nu+(\mathbf{x}-\boldsymbol{\mu})'\mathbf{N}(\mathbf{x}-\boldsymbol{\mu}))^{-(k+\nu)/2}. \tag{A22.1}$$

Proof: Since $\mathbf{z}$ and h are independent, the joint density $p(\mathbf{z},h|\mathbf{N}^{-1},\nu)$ of $\mathbf{z}$ and h is obtained from (A12.3) and (A21.1) by (Koch 1988a, p.107)

$$p(\mathbf{z},h|\mathbf{N}^{-1},\nu) = (2\pi)^{-k/2}(\det\mathbf{N})^{1/2}\exp(-\mathbf{z}'\mathbf{N}\mathbf{z}/2)$$
$$2^{-\nu/2}(\Gamma(\nu/2))^{-1}h^{\nu/2-1}\exp(-h/2).$$

The random vector $\mathbf{z}$ is now transformed by

$$z_i = (h/\nu)^{1/2}(x_i-\mu_i)$$

with

$$\partial z_i/\partial x_i = (h/\nu)^{1/2}.$$

To compute the density function of the transformed variables, the Jacobian $\det\mathbf{J}$ of the transformation is needed (Koch 1988a, p.108). It is determined by

$$\det\mathbf{J} = \prod_{i=1}^{k}(h/\nu)^{1/2} = (h/\nu)^{k/2},$$

so that the density function $p(\mathbf{x},h|\boldsymbol{\mu},\mathbf{N}^{-1},\nu)$ follows with

$$p(\mathbf{x},h|\boldsymbol{\mu},\mathbf{N}^{-1},\nu) = 2^{-(k+\nu)/2}(\nu\pi)^{-k/2}(\Gamma(\nu/2))^{-1}(\det\mathbf{N})^{1/2}$$
$$h^{(k+\nu)/2-1}\exp[-\tfrac{1}{2}(1+(\mathbf{x}-\boldsymbol{\mu})'\mathbf{N}(\mathbf{x}-\boldsymbol{\mu})/\nu)h]. \tag{A22.2}$$

We compute the marginal density of $\mathbf{x}$ (Koch 1988a, p.105)

$$p(\mathbf{x}|\boldsymbol{\mu},\mathbf{N}^{-1},\nu) = 2^{-(k+\nu)/2}(\nu\pi)^{-k/2}(\Gamma(\nu/2))^{-1}(\det\mathbf{N})^{1/2}$$
$$\int_0^\infty h^{(k+\nu)/2-1}\exp(-Qh)dh$$

with

$$Q = \tfrac{1}{2}(1+(\mathbf{x}-\boldsymbol{\mu})'\mathbf{N}(\mathbf{x}-\boldsymbol{\mu})/\nu).$$

From (A12.2) we obtain

$$\int_0^\infty h^{(k+\nu)/2-1}\exp(-Qh)dh = \Gamma((k+\nu)/2)Q^{-(k+\nu)/2}$$

and therefore

$$p(\mathbf{x}|\boldsymbol{\mu},\mathbf{N}^{-1},\nu) = \frac{\Gamma((k+\nu)/2)(\det\mathbf{N})^{1/2}}{(\nu\pi)^{k/2}\Gamma(\nu/2)} (1+(\mathbf{x}-\boldsymbol{\mu})'\mathbf{N}(\mathbf{x}-\boldsymbol{\mu})/\nu)^{-(k+\nu)/2}. \quad \text{(A22.3)}$$

This expression gives, after multiplying numerator and denominator by $\nu^{\nu/2}$, the density function of (A22.1), which completes the proof. □

Example 1: We take the density of the multivariate t-distribution in the form of (A22.3) and set k=1, $\mathbf{x}$=x, $\boldsymbol{\mu}=\mu$, $\mathbf{N}$=f and obtain the density function

$$p(x|\mu,1/f,\nu) = \frac{\Gamma((\nu+1)/2)}{\sqrt{\pi}\ \Gamma(\nu/2)} (\frac{f}{\nu})^{1/2} (1 + \frac{f}{\nu}(x-\mu)^2)^{-(\nu+1)/2}. \quad \text{(A22.4)}$$

This is the density function of a random variable x having the univariate t-distribution $t(\mu,1/f,\nu)$, thus $x \sim t(\mu,1/f,\nu)$. The standardized form of this distribution follows from the transformation of the variable x to z by

$$z = \sqrt{f}(x-\mu). \quad \text{(A22.5)}$$

With $dx/dz=1/\sqrt{f}$ we obtain the density function

$$p(z|\nu) = \frac{\Gamma((\nu+1)/2)}{\sqrt{\nu\pi}\ \Gamma(\nu/2)} (1 + \frac{z^2}{\nu})^{-(\nu+1)/2}, \quad \text{(A22.6)}$$

which is the density function of a random variable z having the t-distribution $t(\nu)$, also called Student's t-distribution, thus $z \sim t(\nu)$ (Koch 1988a, p.154). The distribution (A22.1) is therefore the multivariate generalization of the t-distribution. Δ

The first moment and the second central moment of a random vector having the multivariate t-distribution shall be given next.

Theorem: Let the k×1 random vector $\mathbf{x}$ be distributed according to $\mathbf{x} \sim t(\boldsymbol{\mu},\mathbf{N}^{-1},\nu)$, then

$$E(\mathbf{x}) = \boldsymbol{\mu} \quad \text{for} \quad \nu > 1$$

and

$$D(\mathbf{x}) = \nu(\nu-2)^{-1}\mathbf{N}^{-1} \quad \text{for} \quad \nu > 2. \quad \text{(A22.7)}$$

Proof: We start from the density function (A22.2), whose marginal density gives the multivariate t-distribution. Rearranging leads to

$$p(\mathbf{x},h|\boldsymbol{\mu},\mathbf{N}^{-1},\nu) = 2^{-\nu/2}\,(\Gamma(\nu/2))^{-1}\,h^{\nu/2-1}\,\exp(-h/2)$$

$$(2\pi)^{-k/2}\,(\det\mathbf{N})^{1/2}\,(h/\nu)^{k/2}\,\exp(-\tfrac{1}{2}(h/\nu)(\mathbf{x}-\boldsymbol{\mu})'\mathbf{N}(\mathbf{x}-\boldsymbol{\mu}))$$

$$= p(h)\;p(\mathbf{x}|\boldsymbol{\mu},(\nu/h)\mathbf{N}^{-1}), \qquad \text{(A22.8)}$$

so that the multivariate t-distribution is obtained by computing the marginal density for $\mathbf{x}$

$$p(\mathbf{x}|\boldsymbol{\mu},\mathbf{N}^{-1},\nu) = \int_0^\infty p(h)\;p(\mathbf{x}|\boldsymbol{\mu},(\nu/h)\mathbf{N}^{-1})dh.$$

The expected value $E(\mathbf{x})$ of $\mathbf{x}$ follows with (Koch 1988a, p.111)

$$E(\mathbf{x}) = \int_{-\infty}^\infty \ldots \int_{-\infty}^\infty \int_0^\infty \mathbf{x}\;p(h)\;p(\mathbf{x}|\boldsymbol{\mu},(\nu/h)\mathbf{N}^{-1})dh\;dx_1\ldots dx_k$$

and after changing the sequence of integration

$$E(\mathbf{x}) = \int_0^\infty [\int_{-\infty}^\infty \ldots \int_{-\infty}^\infty \mathbf{x}\;p(\mathbf{x}|\boldsymbol{\mu},(\nu/h)\mathbf{N}^{-1})dx_1\ldots dx_k]\;p(h)\;dh. \qquad \text{(A22.9)}$$

The inner integral represents the expected value of a random variable $\mathbf{x}$ having the normal distribution $\mathbf{x}\sim N(\boldsymbol{\mu},(\nu/h)\mathbf{N}^{-1})$, as can be seen by comparing (A22.8) with (A21.1). Thus, we obtain with (A21.3)

$$E(\mathbf{x}) = \int_0^\infty \boldsymbol{\mu}\;p(h)\;dh$$

and with (A12.2)

$$\int_0^\infty 2^{-\nu/2}(\Gamma(\nu/2))^{-1}\,h^{\nu/2-1}\,e^{-h/2}\,dh = 1,$$

so that $E(\mathbf{x})=\boldsymbol{\mu}$ follows for $\nu>1$ (Zellner 1971, p.385).

To obtain the covariance matrix $D(\mathbf{x})$ of random vector $\mathbf{x}$ with a multivariate t-distribution, we only have to write instead of (A22.9) (Koch 1988a, p.116)

$$D(\mathbf{x}) = \int_0^\infty [\int_{-\infty}^\infty \ldots \int_{-\infty}^\infty (\mathbf{x}-\boldsymbol{\mu})(\mathbf{x}-\boldsymbol{\mu})'p(\mathbf{x}|\boldsymbol{\mu},(\nu/h)\mathbf{N}^{-1})dx_1\ldots dx_k]\;p(h)\;dh.$$

The inner integral represents the covariance matrix of a random variable $\mathbf{x}$ having the normal distribution $\mathbf{x}\sim N(\boldsymbol{\mu},(\nu/h)\mathbf{N}^{-1})$, thus, with (A21.3)

$$D(\mathbf{x}) = \int_0^\infty (\nu/h)\mathbf{N}^{-1}\;p(h)\;dh$$

and with substituting from (A22.8)

$$D(\mathbf{x}) = \mathbf{N}^{-1} \int_0^\infty 2^{-\nu/2} \, (\Gamma(\nu/2))^{-1} \, \nu h^{\nu/2-2} \, \exp(-h/2) \, dh.$$

The integral (A12.2) gives

$$\int_0^\infty 2^{-(\nu/2-1)} (\Gamma(\nu/2-1))^{-1} h^{\nu/2-2} \, \exp(-h/2) \, dh = 1$$

and therefore

$$D(\mathbf{x}) = \mathbf{N}^{-1} 2^{-\nu/2} \, (\Gamma(\nu/2))^{-1} \nu \, 2^{\nu/2-1} \, \Gamma(\nu/2-1).$$

With $\Gamma(\nu/2)=(\nu/2-1)\Gamma(\nu/2-1)$ we finally obtain

$$D(\mathbf{x}) = \nu(\nu-2)^{-1}\mathbf{N}^{-1} \quad \text{for} \quad \nu > 2,$$

which proves the theorem. □

The marginal and the conditional distribution of the multivariate t-distribution follow by the

Theorem: Let the $k\times 1$ vector $\mathbf{x}$ be distributed according to $\mathbf{x} \sim t(\boldsymbol{\mu}, \mathbf{N}^{-1}, \nu)$. If $\mathbf{x}$ is partitioned into the $(k-m)\times 1$ vector $\mathbf{x}_1$ and the $m\times 1$ vector $\mathbf{x}_2$ with $\mathbf{x}=[\mathbf{x}_1', \mathbf{x}_2']'$ and accordingly $\boldsymbol{\mu}=[\boldsymbol{\mu}_1', \boldsymbol{\mu}_2']'$ as well as the matrix $\mathbf{N}$

$$\mathbf{N} = \begin{bmatrix} \mathbf{N}_{11} & \mathbf{N}_{12} \\ \mathbf{N}_{21} & \mathbf{N}_{22} \end{bmatrix} \quad \text{with} \quad \mathbf{N}^{-1} = \begin{bmatrix} \mathbf{I}_{11} & \mathbf{I}_{12} \\ \mathbf{I}_{21} & \mathbf{I}_{22} \end{bmatrix},$$

then the random vector $\mathbf{x}_2$ has a multivariate t-distribution, too

$$\mathbf{x}_2 \sim t(\boldsymbol{\mu}_2, \mathbf{N}_{22.1}^{-1}, \nu)$$

with the marginal density function

$$p(\mathbf{x}_2|\boldsymbol{\mu}_2, \mathbf{N}_{22.1}^{-1}, \nu) = \frac{\nu^{\nu/2}\Gamma((m+\nu)/2)(\det\mathbf{N}_{22.1})^{1/2}}{\pi^{m/2}\Gamma(\nu/2)}$$

$$(\nu+(\mathbf{x}_2-\boldsymbol{\mu}_2)'\mathbf{N}_{22.1}(\mathbf{x}_2-\boldsymbol{\mu}_2))^{-(m+\nu)/2}, \tag{A22.10}$$

where

$$\mathbf{N}_{22.1} = \mathbf{N}_{22}-\mathbf{N}_{21}\mathbf{N}_{11}^{-1}\mathbf{N}_{12} \quad \text{or} \quad \mathbf{N}_{22.1}^{-1} = \mathbf{I}_{22}.$$

The distribution of the random vector $\mathbf{x}_1$ under the condition that the second random vector takes on the values $\mathbf{x}_2$ has also the form of a multivariate t-distribution with the conditional density function

$$p(\mathbf{x}_1|\mathbf{x}_2,\boldsymbol{\mu}_{1.2},\mathbf{N}_{1.2}^{-1},\nu,m+\nu) = \frac{\nu^{(m+\nu)/2}\Gamma(((k-m)+(m+\nu))/2)(\det\mathbf{N}_{1.2})^{1/2}}{\pi^{(k-m)/2}\Gamma((m+\nu)/2)}$$

$$(\nu+(\mathbf{x}_1-\boldsymbol{\mu}_{1.2})'\mathbf{N}_{1.2}(\mathbf{x}_1-\boldsymbol{\mu}_{1.2}))^{-((k-m)+(m+\nu))/2},$$

where

$$\boldsymbol{\mu}_{1.2} = \boldsymbol{\mu}_1 - \mathbf{N}_{11}^{-1}\mathbf{N}_{12}(\mathbf{x}_2-\boldsymbol{\mu}_2)$$

and

$$\mathbf{N}_{1.2} = \mathbf{N}_{11}/(1+(\mathbf{x}_2-\boldsymbol{\mu}_2)'\mathbf{N}_{22.1}(\mathbf{x}_2-\boldsymbol{\mu}_2)/\nu). \qquad \text{(A22.11)}$$

Proof: To derive the marginal density function for $\mathbf{x}_2$ we have to integrate the density function of the multivariate t-distribution with respect to $\mathbf{x}_1$. We take the density in the form of (A22.3), where for the sake of simplification we substitute $\mathbf{N}/\nu=\mathbf{M}$ with $\nu>0$ and obtain

$$p(\mathbf{x}|\boldsymbol{\mu},(\nu\mathbf{M})^{-1},\nu) = \frac{\Gamma((k+\nu)/2)(\det\mathbf{M})^{1/2}}{\pi^{k/2}\Gamma(\nu/2)}\,(1+(\mathbf{x}-\boldsymbol{\mu})'\mathbf{M}(\mathbf{x}-\boldsymbol{\mu}))^{-(k+\nu)/2}.$$

By substituting the partitioning of (A22.10) we rewrite the quadratic form

$$\begin{aligned}(\mathbf{x}-\boldsymbol{\mu})'\mathbf{M}(\mathbf{x}-\boldsymbol{\mu}) &= (\mathbf{x}_1-\boldsymbol{\mu}_1)'\mathbf{M}_{11}(\mathbf{x}_1-\boldsymbol{\mu}_1) + 2(\mathbf{x}_1-\boldsymbol{\mu}_1)'\mathbf{M}_{12}(\mathbf{x}_2-\boldsymbol{\mu}_2)\\ &\quad + (\mathbf{x}_2-\boldsymbol{\mu}_2)'\mathbf{M}_{22}(\mathbf{x}_2-\boldsymbol{\mu}_2)\\ &= (\mathbf{x}_1-\boldsymbol{\mu}_1+\mathbf{M}_{11}^{-1}\mathbf{M}_{12}(\mathbf{x}_2-\boldsymbol{\mu}_2))'\mathbf{M}_{11}(\mathbf{x}_1-\boldsymbol{\mu}_1+\mathbf{M}_{11}^{-1}\mathbf{M}_{12}(\mathbf{x}_2-\boldsymbol{\mu}_2))\\ &\quad + (\mathbf{x}_2-\boldsymbol{\mu}_2)'\mathbf{M}_{22.1}(\mathbf{x}_2-\boldsymbol{\mu}_2)\\ &= Q_{1.2} + Q_2\end{aligned}$$

with

$$\mathbf{M}_{22.1} = \mathbf{M}_{22}-\mathbf{M}_{21}\mathbf{M}_{11}^{-1}\mathbf{M}_{12}.$$

The determinant of the block matrix $\mathbf{M}$ is computed by (Koch 1988a, p.45)

$$\det\mathbf{M} = \det\mathbf{M}_{11}\,\det\mathbf{M}_{22.1}.$$

Substituting these results into the density function leads to

$$p(\mathbf{x}_1,\mathbf{x}_2|\boldsymbol{\mu},(\nu\mathbf{M})^{-1},\nu) = \frac{\Gamma((m+\nu)/2)(\det\mathbf{M}_{22.1})^{1/2}}{\pi^{m/2}\Gamma(\nu/2)}(1+Q_2)^{-(m+\nu)/2}$$

$$\frac{\Gamma((k+\nu)/2)(\det\mathbf{M}_{11})^{1/2}}{\pi^{(k-m)/2}\Gamma((m+\nu)/2)}(1+Q_2)^{-(k-m)/2}(1+Q_{1.2}/(1+Q_2))^{-(k+\nu)/2}.$$

We now substitute $\mathbf{M}=\mathbf{N}/\nu$ and obtain after multiplying numerator and denominator of the first factor by $\nu^{\nu/2}$ and of the second factor by $\nu^{(m+\nu)/2}$

$$p(\mathbf{x}_1,\mathbf{x}_2|\boldsymbol{\mu},\mathbf{N}^{-1},\nu) = \frac{\nu^{\nu/2}\Gamma((m+\nu)/2)(\det\mathbf{N}_{22.1})^{1/2}}{\pi^{m/2}\Gamma(\nu/2)}(\nu+\nu Q_2)^{-(m+\nu)/2}$$

$$\frac{\nu^{(m+\nu)/2}\Gamma((k+\nu)/2)[\det(\mathbf{N}_{11}/(1+Q_2))]^{1/2}}{\pi^{(k-m)/2}\Gamma((m+\nu)/2)}(\nu+\nu Q_{1.2}/(1+Q_2))^{-(k+\nu)/2}.$$

The first factor gives the marginal density function of $\mathbf{x}_2$ and the second factor the conditional density function of $\mathbf{x}_1$ given $\mathbf{x}_2$ because of

$$p(\mathbf{x}_1,\mathbf{x}_2|\boldsymbol{\mu},\mathbf{N}^{-1},\nu) = p(\mathbf{x}_2|\boldsymbol{\mu},\mathbf{N}^{-1},\nu)p(\mathbf{x}_1|\mathbf{x}_2,\boldsymbol{\mu},\mathbf{N}^{-1},\nu,m+\nu)$$

from (261.1). With the inverse of a block matrix (Koch 1988a, p.39) we obtain $\mathbf{N}_{22.1}^{-1}=\mathbf{I}_{22}$. This proves the theorem. □

The distribution of a random vector which originates from a linear transformation of a random vector with the multivariate t-distribution is derived next.

Theorem: Let the k×1 vector $\mathbf{x}$ be distributed according to $\mathbf{x}\sim t(\boldsymbol{\mu},\mathbf{N}^{-1},\nu)$, then the m×1 random vector $\mathbf{y}$ obtained by the linear transformation $\mathbf{y}=\mathbf{A}\mathbf{x}+\mathbf{c}$, where $\mathbf{A}$ denotes an m×k matrix of constants with full row rank and $\mathbf{c}$ an m×1 vector of constants, has the multivariate t-distribution

$$\mathbf{y} \sim t(\mathbf{A}\boldsymbol{\mu}+\mathbf{c},\mathbf{A}\mathbf{N}^{-1}\mathbf{A}',\nu)$$

with the density function

$$p(\mathbf{y}|\mathbf{A}\boldsymbol{\mu}+\mathbf{c},\mathbf{A}\mathbf{N}^{-1}\mathbf{A}',\nu) = \frac{\nu^{\nu/2}\Gamma((m+\nu)/2)(\det\mathbf{A}\mathbf{N}^{-1}\mathbf{A}')^{-1/2}}{\pi^{m/2}\Gamma(\nu/2)}$$

$$(\nu+(\mathbf{y}-\mathbf{A}\boldsymbol{\mu}-\mathbf{c})'(\mathbf{A}\mathbf{N}^{-1}\mathbf{A}')^{-1}(\mathbf{y}-\mathbf{A}\boldsymbol{\mu}-\mathbf{c}))^{-(m+\nu)/2}. \qquad (A22.12)$$

Proof: We have to distinguish two cases,

a) $m = k$.

Since $\mathbf{A}$ is assumed to be of full row rank, $\mathbf{A}^{-1}$ exists and

$$\mathbf{x} = \mathbf{A}^{-1}(\mathbf{y}-\mathbf{c}) .$$

The Jacobian of this transformation is given by $\det\mathbf{J}=\det\mathbf{A}^{-1}$ (Koch 1988a, p.85). Introducing it together with $\mathbf{x}-\boldsymbol{\mu}=\mathbf{A}^{-1}(\mathbf{y}-\mathbf{A}\boldsymbol{\mu}-\mathbf{c})$ into the density function of (A22.1) gives the density (A22.12) because of $\det\mathbf{A}^{-1}(\det\mathbf{N})^{1/2}=(\det\mathbf{A}\mathbf{N}^{-1}\mathbf{A}')^{-1/2}$.

b) $m < k$.

Because of the full row rank of the $m\times k$ matrix $\mathbf{A}$ we may add $k-m$ linearly independent rows to $\mathbf{A}$, collected in the matrix $\mathbf{B}$, so that the regular matrix $\mathbf{C}$ with $\mathbf{C}=[\mathbf{A}',\mathbf{B}']'$ is obtained. If the vector $\mathbf{c}$ is also augmented by a $(k-m)\times 1$ vector $\mathbf{d}$ of constants, so that $\mathbf{e}=[\mathbf{c}',\mathbf{d}']'$ follows, we find by the linear transformation the random vector $\mathbf{u}=[\mathbf{y}',\mathbf{z}']'$ having according to case a) the multivariate t-distribution

$$\mathbf{u} \sim t(\mathbf{C}\boldsymbol{\mu}+\mathbf{e},\mathbf{C}\mathbf{N}^{-1}\mathbf{C}',\nu) .$$

The marginal distribution of $\mathbf{y}$ has to be determined, whose parameters according to (A22.10) follow with $\mathbf{A}\boldsymbol{\mu}+\mathbf{c}$ and the first block matrix on the diagonal of the matrix $\mathbf{C}\mathbf{N}^{-1}\mathbf{C}'$, which is $\mathbf{A}\mathbf{N}^{-1}\mathbf{A}'$. But this gives the density (A22.12), so that the theorem is proved. □

Finally we will present the relation between the multivariate t-distribution and the F-distribution.

Theorem: Let the $k\times 1$ vector $\mathbf{x}$ be distributed according to $\mathbf{x}\sim t(\boldsymbol{\mu},\mathbf{N}^{-1},\nu)$, then the quadratic form $(\mathbf{x}-\boldsymbol{\mu})'\mathbf{N}(\mathbf{x}-\boldsymbol{\mu})/k$ has the F-distribution $F(k,\nu)$ with k and ν degrees of freedom

$$(\mathbf{x}-\boldsymbol{\mu})'\mathbf{N}(\mathbf{x}-\boldsymbol{\mu})/k \sim F(k,\nu) . \qquad \text{(A22.13)}$$

Proof: We will apply the transformation

$$\mathbf{y} = \mathbf{A}^{-1}(\mathbf{x}-\boldsymbol{\mu}) ,$$

where $\mathbf{A}$ is a regular $k\times k$ matrix with the property $\mathbf{A}'\mathbf{N}\mathbf{A}=\mathbf{I}$, so that $\mathbf{N}^{-1}=\mathbf{A}\mathbf{A}'$ holds. Then the density of the multivariate t-distribution is transformed because of (A22.12) to the standardized form

$$p(\mathbf{y}|\mathbf{0},\mathbf{I},\nu) = \frac{\nu^{\nu/2}\Gamma((k+\nu)/2)}{\pi^{k/2}\Gamma(\nu/2)} (\nu+\mathbf{y}'\mathbf{y})^{-(k+\nu)/2}. \tag{A22.14}$$

We make a transformation to polar coordinates with

$$y_1 = \sqrt{r}\ \cos\alpha_1\cos\alpha_2\ \ldots\ \cos\alpha_{k-1}$$

$$y_2 = \sqrt{r}\ \cos\alpha_1\cos\alpha_2\ \ldots\ \cos\alpha_{k-2}\sin\alpha_{k-1}$$

$$\ldots$$

$$y_j = \sqrt{r}\ \cos\alpha_1\cos\alpha_2\ \ldots\ \cos\alpha_{k-j}\sin\alpha_{k-j+1}$$

$$\ldots$$

$$y_k = \sqrt{r}\ \sin\alpha_1,$$

where

$$0 \le r < \infty,\ 0 \le \alpha_{k-1} < 2\pi,\ -\pi/2 \le \alpha_i \le \pi/2 \quad \text{for} \quad i\in\{1,\ldots,k-2\},$$

and obtain

$$r = \mathbf{y}'\mathbf{y} = y_1^2 + y_2^2 + \ldots + y_k^2.$$

The Jacobian of this transformation follows from the kxk matrix $\mathbf{J}$, whose jth row contains the elements $\partial y_j/\partial r$, $\partial y_j/\partial\alpha_1,\ldots,\partial y_j/\partial\alpha_{k-1}$. Furthermore, by multiplication we find

$$\mathbf{J}'\mathbf{J} = \mathrm{diag}(d_1,d_2,\ldots,d_k)$$

$$= \mathrm{diag}((4r)^{-1},r,r\cos^2\alpha_1,r\cos^2\alpha_1\cos^2\alpha_2,\ldots,r\cos^2\alpha_1\ldots\cos^2\alpha_{k-2}),$$

therefore

$$\det(\mathbf{J}'\mathbf{J}) = \prod_{i=1}^{k} d_i \quad \text{or} \quad \det\mathbf{J} = \prod_{i=1}^{k} \sqrt{d_i}$$

and finally

$$\det\mathbf{J} = (1/2)r^{k/2-1}\cos^{k-2}\alpha_1\cos^{k-3}\alpha_2\ \ldots\ \cos\alpha_{k-2}.$$

Substituting these results gives instead of (A22.14)

$$p(r,\alpha_1,\ldots,\alpha_{k-1}|k,\nu) = \det\mathbf{J}\ \frac{\nu^{\nu/2}\Gamma((k+\nu)/2)}{\pi^{k/2}\Gamma(\nu/2)} (\nu+r)^{-(k+\nu)/2}.$$

The marginal distribution of r is found by integrating over α_i. We start from the integral (Gradshteyn and Ryzhik 1965, p.369)

$$\int_0^{\pi/2} \cos^{k-j-1}\alpha d\alpha = 2^{k-j-2}\,\Gamma(\tfrac{k-j}{2})\Gamma(\tfrac{k-j}{2})/\Gamma(k-j).$$

In addition we have

$$\int_{-\pi/2}^{\pi/2} \cos^{k-j-1}\alpha d\alpha = 2\int_0^{\pi/2} \cos^{k-j-1}\alpha d\alpha,$$

and using the recursion formula of the gamma function we find with $\Gamma(1/2)=\sqrt{\pi}$

$$\Gamma(\tfrac{k-j+1}{2})\ \Gamma(\tfrac{k-j}{2}) = \sqrt{\pi}\,\Gamma(k-j)/2^{k-j-1},$$

so that we compute the integrals

$$\int_{-\pi/2}^{\pi/2} \cos^{k-j-1}\alpha_j d\alpha_j = \sqrt{\pi}\,\Gamma(\tfrac{k-j}{2})/\Gamma(\tfrac{k-j+1}{2}) \quad \text{for} \quad j\in\{1,\ldots,k-2\}$$

and

$$\int_0^{2\pi} d\alpha_{k-1} = 2\pi.$$

These results give the density

$$p(r|k,\nu) = \frac{\pi^{(k-2)/2}\nu^{\nu/2}\Gamma((k+\nu)/2)}{\pi^{-1}\pi^{k/2}\Gamma(k/2)\Gamma(\nu/2)}\, r^{k/2-1}\,(\nu+r)^{-(k+\nu)/2}.$$

We finally transform from r to w with $r=kw$ and $dr/dw=k$, so that the density is obtained

$$p(w|k,\nu) = \frac{\Gamma((k+\nu)/2)k^{k/2}\nu^{\nu/2}w^{k/2-1}}{\Gamma(k/2)\Gamma(\nu/2)(\nu+kw)^{(k+\nu)/2}} \quad \text{for} \quad 0 < w < \infty.$$

This is the density function of the F-distribution $F(k,\nu)$ with k and ν degrees of freedom (Koch 1988a, p.149). Because of $w=\mathbf{y}'\mathbf{y}/k=(\mathbf{x}-\boldsymbol{\mu})'\mathbf{N}(\mathbf{x}-\boldsymbol{\mu})/k$ the theorem is proved. □

A23 Normal-Gamma Distribution

Theorem: Let $\mathbf{x}$ be an $n\times 1$ random vector with $\mathbf{x}=(x_i)$ and τ a random variable. Let the distribution of $\mathbf{x}$ under the condition that τ takes on the value τ be given by the normal distribution $N(\boldsymbol{\mu},\tau^{-1}\mathbf{V})$ with the $n\times 1$ vector $\boldsymbol{\mu}$ and the positive definite $n\times n$ matrix $\tau^{-1}\mathbf{V}$ as parameters, hence $\mathbf{x}|\tau\sim N(\boldsymbol{\mu},\tau^{-1}\mathbf{V})$, and let the random variable τ have the gamma distribution $G(b,p)$ with b and p as parameters, thus $\tau\sim G(b,p)$. Then the random vector

$z=[x',\tau]'$ is said to have the *normal-gamma* distribution $NG(\boldsymbol{\mu},\mathbf{V},b,p)$ with the parameters $\boldsymbol{\mu},\mathbf{V},b,p$, hence $\mathbf{x},\tau \sim NG(\boldsymbol{\mu},\mathbf{V},b,p)$, and the density function given by

$$p(\mathbf{x},\tau|\boldsymbol{\mu},\mathbf{V},b,p) = (2\pi)^{-n/2}(\det\mathbf{V})^{-1/2}b^p(\Gamma(p))^{-1}$$
$$\tau^{n/2+p-1}\exp\{-\frac{\tau}{2}[2b+(\mathbf{x}-\boldsymbol{\mu})'\mathbf{V}^{-1}(\mathbf{x}-\boldsymbol{\mu})]\}$$

for $b>0,\ p>0,\ 0<\tau<\infty,\ -\infty<x_i<\infty.$ (A23.1)

Proof: The conditional density function $p(\mathbf{x}|\boldsymbol{\mu},\tau^{-1}\mathbf{V})$ of $\mathbf{x}$ given τ is obtained by the density (A21.1) of the multivariate normal distribution and the density function $p(\tau|b,p)$ of τ by the density (A12.1) of the gamma distribution. The joint density of $\mathbf{x}$ and τ follows from (211.2) with

$$p(\mathbf{x},\tau|\boldsymbol{\mu},\mathbf{V},b,p) = p(\mathbf{x}|\boldsymbol{\mu},\tau^{-1}\mathbf{V})p(\tau|b,p), \tag{A23.2}$$

which immediately leads to the density function (A23.1) of the normal-gamma distribution. □

The marginal distributions of $\mathbf{x}$ and τ shall be given next.

Theorem: Let the vector $\mathbf{z}$ with $z=[x',\tau]'$ have the normal gamma distribution $\mathbf{x},\tau\sim NG(\boldsymbol{\mu},\mathbf{V},b,p)$, then the marginal distribution of $\mathbf{x}$ is the multivariate t-distribution

$$\mathbf{x} \sim t(\boldsymbol{\mu},b\mathbf{V}/p,2p) \tag{A23.3}$$

and the marginal distribution of τ is the gamma distribution

$$\tau \sim G(b,p). \tag{A23.4}$$

Proof: To obtain the marginal distribution of $\mathbf{x}$ the variable τ has to be integrated out of the density function of (A23.1). We find with (A12.2)

$$\int_0^\infty \tau^{n/2+p-1}\exp\{-\frac{\tau}{2}[2b+(\mathbf{x}-\boldsymbol{\mu})'\mathbf{V}^{-1}(\mathbf{x}-\boldsymbol{\mu})]\}d\tau$$
$$= \Gamma(n/2+p)\{\frac{1}{2}[2b+(\mathbf{x}-\boldsymbol{\mu})'\mathbf{V}^{-1}(\mathbf{x}-\boldsymbol{\mu})]\}^{-n/2-p}.$$

Substituting this result in (A23.1) gives

$$p(\mathbf{x}|\boldsymbol{\mu},b\mathbf{V}/p,2p) = (2b)^p\ (\pi)^{-n/2}(\det\mathbf{V})^{-1/2}(\Gamma(p))^{-1}$$
$$\Gamma(n/2+p)(2b+(\mathbf{x}-\boldsymbol{\mu})'\mathbf{V}^{-1}(\mathbf{x}-\boldsymbol{\mu}))^{-n/2-p}$$

$$= \frac{\Gamma((n+2p)/2)(\det 2p\mathbf{V}^{-1}/2b)^{1/2}}{(2p\pi)^{n/2}\Gamma(p)}\,(1+(\mathbf{x}-\boldsymbol{\mu})'(2p\mathbf{V}^{-1}/2b)(\mathbf{x}-\boldsymbol{\mu})/2p)^{-(n+2p)/2}.$$

This is, according to (A22.3), the density function of the multivariate t-distribution $t(\mathbf{x}|\boldsymbol{\mu}, b\mathbf{V}/p, 2p)$.

The density of the normal-gamma distribution is obtained by (A23.2), where $p(\mathbf{x}|\boldsymbol{\mu}, \tau^{-1}\mathbf{V})$ is the normal distribution. Hence, integrating with respect to $\mathbf{x}$ gives $p(\tau|b,p)$ as the marginal distribution for τ, which proves the theorem. □

References

Aarts EHL, Laarhoven PJM van (1987) Simulated annealing: a pedestrian review of the theory and some applications. In: Devijver PA, Kittler J (eds) Pattern recognition theory and applications, Springer, Berlin Heidelberg New York Tokyo, pp 179-192

Abramowitz M, Stegun IA (ed) (1965) Handbook of mathematical functions. Dover, New York

Akaike H (1979) A Bayesian extension of the minimum AIC procedure of autoregressive model fitting. Biometrika, 66: 237-242

Berger JO (1985) Statistical decision theory and Bayesian analysis. Springer, Berlin Heidelberg New York Tokyo

Berger JO, Sellke T (1987) Testing a point null hypothesis: the irreconcilability of P values and evidence. J Am Stat Assoc 82: 112-122

Bosch K (1985) Elementare Einführung in die angewandte Statistik. Vieweg, Braunschweig

Bossler JD (1972) Bayesian inference in geodesy. Dissertation, The Ohio State University, Columbus

Bossler JD, Hanson RH (1980) Application of special variance estimators to geodesy. NOAA Technical Report NOS 84 NGS 15, US Department of Commerce, National Geodetic Survey, Rockville

Box GEP, Jenkins GM (1970) Time series analysis. Holden-Day, San Francisco

Box GEP, Tiao GC (1973) Bayesian inference in statistical analysis. Addison-Wesley, Reading

Broemeling LD (1985) Bayesian analysis of linear models. Dekker, New York

Bunke H, Bunke O (ed) (1986) Statistical inference in linear models, vol.I. Wiley, New York

Busch A, Koch KR (1990) Reconstruction of digital images using Bayesian estimates. Z Photogrammetrie und Fernerkundung, 58, in print

Casella G, Berger RL (1987) Reconciling Bayesian and frequentist evidence in the one-sided testing problem. J Am Stat Assoc 82: 106-111

Caspary W (1988) Fehlerverteilungen, Methode der kleinsten Quadrate und robuste Alternativen. Z Vermessungswes 113: 123-133

Cramer H (1946) Mathematical methods of statistics. Princeton University Press, Princeton

DeGroot MH (1970) Optimal statistical decisions. McGraw-Hill, New York

Dillinger WH, Pope AJ, Harding ST (1971) The determination of focal mechanisms using P- and S-wave data. NOAA Technical Report NOS 44, US Department of Commerce, National Ocean Survey, Rockville

Dowson DC, Wragg A (1973) Maximum-entropy distributions having prescribed first and second moments. IEEE Transactions on Information Theory, IT-19: 689-693

Eeg J, Krarup T (1973) Integrated geodesy. The Danish Geodetic Institute, Int Rep 7, Kobenhavn

Förstner W (1988) Statistische Verfahren für die automatische Bildanalyse und ihre Bewertung bei der Objekterkennung und -vermessung. Habilitationsschrift, Institut für Photogrammetrie der Universität Stuttgart, Stuttgart

Frühwirth R, Regler M (1983) Monte-Carlo-Methoden. Bibliograph Inst, Mannheim

Geisser S (1964) Posterior odds for multivariate normal classifications. J Royal Stat Soc, Ser B, 26: 69-76

Geisser S (1965) Bayesian estimation in multivariate analysis. Ann Math Statist 36: 150-159

Geman D, Geman S, Graffigne C (1987) Locating texture and object boundaries. In: Devijver PA, Kittler J (eds) Pattern recognition theory and applications, Springer, Berlin Heidelberg New York Tokyo, pp 165-177

Geman S, Geman D (1984) Stochastic relaxation, Gibbs distributions, and Bayesian restoration of images. IEEE Trans Pattern Anal Machine Intell, PAMI-6: 721-741

Gradshteyn IS, Ryzhik IM (1965) Table of integrals, series, and products. Academic Press, New York London

Guiasu S (1977) Information theory with applications. McGraw-Hill, New York

Gull SF, Skilling J (1985) The entropy of an image. In: Smith CR, Grandy WT (eds) Maximum-entropy and Bayesian methods in inverse problems, Reidel, Dodrecht, pp 287-301

Hammersley JM, Handscomb DC (1964) Monte Carlo methods. Methuen, London

Hampel FR (1974) The influence curve and its role in robust estimation. J Amer Statist Assoc 69: 383-393

Hampel FR, Ronchetti EM, Rousseeuw PJ, Stahel WA (1986) Robust statistics. Wiley, New York

Harvey BR (1987) Degrees of freedom-simplified. Aust J Geod Photogram Surv 46 and 47: 57-68

Hein GW (1986) Integrated geodesy, state-of-the-art 1986 reference text. In: Sünkel H (ed) Lecture Notes in Earth Sciences, vol.7, Mathematical and numerical techniques in physical geodesy, Springer, Berlin Heidelberg New York Tokyo, pp 505-548

Hoaglin DC, Mosteller F, Tukey JW (1983) Understanding robust and exploratory data analysis. Wiley, New York

Huber PJ (1977) Robust methods of estimation of regression equations. Math Operationsforsch Statist, Ser. Stat 8: 141-153

Huber PJ (1981) Robust statistics. Wiley, New York

Jaynes ET (1986) Bayesian methods: general background. In: Justice JH (ed) Maximum entropy and Bayesian methods in applied statistics. Cambridge Univ Press, Cambrigde, pp 1-25

Jazwinski AH (1970) Stochastic processes and filtering theory. Academic Press, New York London

Jeffreys H (1961) Theory of probability. Clarendon, Oxford

Jorgensen PC, Kubik K, Weng PFW (1985) Ah, robust estimation. Aust J Geod Photogram Surv 42: 19-32

Koch KR (1982) Kalman filter and optimal smoothing derived by the regression model. Manuscripta Geodaetica 7: 133-144

Koch KR (1984) Statistical tests for detecting crustal movements using Bayesian inference. NOAA Technical Report NOS NGS 29, US Department of Commerce, National Geodetic Survey, Rockville

Koch KR (1985) Ein statistisches Auswerteverfahren für Deformationsmessungen. Allg Vermess Nachr 92: 97-108

Koch KR (1986) Maximum likelihood estimate of variance components, ideas by A.J. Pope. Bull Geod 60: 329-338

Koch KR (1987) Bayesian inference for variance components. Manuscripta Geodaetica 12: 309-313

Koch KR (1988a) Parameter estimation and hypothesis testing in linear models. Springer, Berlin Heidelberg New York Tokyo

Koch KR (1988b) Bayesian statistics for variance components with informative and noninformative priors. Manuscripta Geodaetica 13: 370-373

Koch KR (1988c) Konfidenzintervalle der Bayes-Statistik für die Varianzen von Streckenmessungen auf Eichlinien. Vermessung Photogrammetrie Kulturtechnik 86: 337-340

Koch KR (1989) Bayes-Statistik mittels Monte-Carlo-Integration. Z Vermessungswes 114: 302-310

Koch KR, Riesmeier K (1985) Bayesian inference for the derivation of less sensitive hypothesis tests. Bull Geod 59: 167-179

Kubik K (1970) The estimation of the weights of measured quantities within the method of least squares. Bull Geod 95: 21-40

Lindley DV (1957) A statistical paradox. Biometrika 44: 187-192

Lindley DV (1965) Introduction to probability and statistics, Part 1 and Part 2. Cambridge Univ Press, Cambridge

Moritz H (1980) Advanced physical geodesy. Wichmann, Karlsruhe

Pilz J (1983) Bayesian estimation and experimental design in linear regression models. Teubner, Leipzig

Press SJ (1982) Applied multivariate analysis. Krieger, Malabar

Press SJ (1989) Bayesian statistics: principles, models, and applications. Wiley, New York

Raiffa H, Schlaifer R (1961) Applied statistical decision theory. Graduate School of Business Administration, Harvard University, Boston

Rao CR (1973) Linear statistical inference and its applications. Wiley, New York

Riesmeier K (1984) Test von Ungleichungshypothesen in linearen Modellen mit Bayes-Verfahren. Dtsch Geod Komm C, 292, München

Ripley BD (1988) Statistical inference for spatial processes. Cambridge Univ Press, Cambridge

Rubinstein RY (1981) Simulation and the Monte Carlo method. Wiley, New York

Schaffrin B (1987) Approximating the Bayesian estimate of the standard deviation in a linear model. Bull Geod 61: 276-280

Schaffrin B (1989) An alternative approach to robust collocation. Bull Geod 63: 395-404

Schwarz KP (1983) Inertial surveying and geodesy. Rev Geophys Space Phys 21: 878-890

Shafer G (1982) Lindley's paradox. J Am Stat Assoc 77: 325-334

Skilling J, Gull SF (1985) Algorithms and applications. In: Smith CR, Grandy WT (eds) Maximum-entropy and Bayesian methods in inverse problems. Reidel, Dodrecht, pp 83-132

Somogyi J (1988) Robust estimation and their use in geodesy. Acta Geod Geoph Mont Hung 23: 45-53

Spitzer F (1971) Markov random fields and Gibbs ensembles. Am Math Mon 78: 142-154

Stearns SD (1975) Digital signal analysis. Hayden, Rochelle Park

Teunissen PJG (1987) The 1 and 2D symmetric Helmert transformation: an exact non-linear least-squares solution. Reports of the Faculty of Geodesy, Delft University of Technology, Delft

Teunissen PJG (1988) The non-linear 2D symmetric Helmert transformation: an exact non-linear least-squares solution. Bull Geod 62: 1-15

Theil H (1963) On the use of incomplete prior information in regression analysis. J Am Stat Assoc 58: 401-414

Ulrych TJ, Bishop TN (1975) Maximum entropy spectral analysis and autoregressive decomposition. Rev Geophys Space Phys 13: 183-200

Wei M (1987) Statistical problems in collocation. Manuscripta Geodaetica 12: 282-289

West M, Harrison J (1989) Bayesian forecasting and dynamic models. Springer, Berlin Heidelberg New York Tokyo

Wolf H (1968) Ausgleichungsrechnung nach der Methode der kleinsten Quadrate. Dümmler, Bonn

Zellner A (1971) An introduction to Bayesian inference in econometrics. Wiley, New York

Ziqiang O (1990) Bayes and emperical Bayes estimation for variance components. Allg Vermess Nachr-Internat Ed 7: 25-31

Index

Lecture Notes in Earth Sciences